Series Editors' Foreword

Oxford Chemistry Masters are designed to provide clear and concise accounts of important topics – both established and emergent – that may be encountered by chemistry students as they progress through postgraduate studies to leadership in research. These Masters assume little prior knowledge other than the foundations provided by an undergraduate degree course in chemistry and lead the reader through an appreciation of the state-of-the-art in the topic whilst providing an entrée to the original literature in the field.

Monosaccharides and oligosaccharides are no longer a specialized branch of chemistry but have moved into the mainstream of chemistry research now that their important role in biological systems is recognized. In this Master Helen Osborn and Tariq Khan have produced an authoritative yet easy to read introduction to the synthesis and biological roles of oligosaccharides. The book will be of interest to all research chemists whatever the stage of their career, be they apprentice or master.

Richard G. Compton
Stephen G. Davies
John Evans

Preface

Exciting developments in analytical methods have recently been reported which allow isolation and identification of previously elusive carbohydrates. As a result carbohydrates have received considerable attention from both biologists and chemists who are attempting to fully understand their biological roles. In this book we wish to portray some of the current excitement in this rapidly growing area of research and highlight some of the challenges facing carbohydrate chemists and biologists. We wish to discuss a number of modern approaches for synthesising carbohydrates so that readers of this book will feel able to contribute to this area of research. The therapeutic potential of carbohydrates will also be discussed to further illustrate their importance.

This book is intended for use by anyone new to carbohydrate chemistry. A reasonable knowledge of organic chemistry is, however, assumed. In particular this book should prove invaluable to final year MChem students as well as MSc and PhD research students.

We would like to thank the following colleagues for their encouragement in preparing this book and for proof-reading the text: Professor John Mann, Victoria Brome, Natasha Gemmell, Jonathan Gridley, Dr Allan Jordan, Hugh Malkin, Sadie Osborne, Will Suthers and Joanne Tait. HMIO is also grateful to Professor Steve Ley, FRS, for introducing her to this exciting area of research.

Reading   H.M.I.O.
Salisbury   T.H.K.
January 2000

To Alixander Sikander, and all who may follow . . .

OXFORD CHEMISTRY MASTERS

OXFORD CHEMISTRY MASTERS

Series Editors

RICHARD G. COMPTON
University of Oxford

STEPHEN G. DAVIES
University of Oxford

JOHN EVANS
University of Southampton

Oligosaccharides:

Their synthesis and biological roles

Helen M.I. Osborn
University of Reading

and

Tariq H. Khan
Enzacta Ltd

OXFORD
UNIVERSITY PRESS

OXFORD

UNIVERSITY PRESS

Great Clarendon Street, Oxford OX2 6DP

Oxford University Press is a department of the University of Oxford.
It furthers the University's objective of excellence in research, scholarship,
and education by publishing worldwide in

Oxford New York

Athens Auckland Bangkok Bogotá Buenos Aires Calcutta
Cape Town Chennai Dar es Salaam Delhi Florence Hong Kong Istanbul
Karachi Kuala Lumpur Madrid Melbourne Mexico City Mumbai
Nairobi Paris São Paulo Singapore Taipei Tokyo Toronto Warsaw

with associated companies in Berlin Ibadan

Oxford is a registered trade mark of Oxford University Press
in the UK and in certain other countries

Published in the United States
by Oxford University Press Inc., New York

© Helen M.I. Osborn and Tariq H. Khan, 2000

British Library Cataloguing in Publication Data
Data available

Library of Congress Cataloging in Publication Data
Osborn, Helen M.I.
Oligosaccharides their chemistry and biological roles / Helen M.I. Osborn
and Tariq H. Khan.
(Oxford chemistry masters ; 3)
Includes bibliographical references and index.
1. Oligosaccharides. I. Khan, Tariq H. II. Title. III. Series.
QD321.O785 2000   572'.565—dc21   99–086892

ISBN 0 19 850260 5 (pbk)
0 19 850265 6 (hbk)

Typeset by J&L Composition Ltd, Filey, N. Yorks
Printed in Great Britain
on acid-free paper by
Bath Press Ltd, Bath, Avon

Contents

1 Introduction

Spectacular advances in all disciplines of science have been realised in recent years and our understanding of many biological processes has continued to improve as the full implications of new discoveries have become appreciated. However, many questions which probe the depth of this understanding remain unanswered. For example, some questions of therapeutic importance include: why are some tumour cells able to migrate from their original cancerous site and produce further tumours at secondary sites? Are there any unusual properties of the Human Immuno-deficiency Virus (HIV) which explain why AIDS is so difficult to treat? How do bacteria, viruses and toxins evade the body's immune system and initiate infective processes? If we could answer questions such as these we would undoubtedly improve our ability to treat such life-threatening diseases, which have defied effective therapy for many years. Many scientists now believe that carbohydrates may provide the missing clues to these puzzling questions. Over the past few years there has been a resurgence of interest in carbohydrate chemistry and biochemistry, and this reflects the important roles which carbohydrates play within biological systems. Thus carbohydrates which have traditionally been merely associated with energy storage and skeletal components are now being associated with a wider range of biological processes within all living systems.

Location of oligosaccharides

Carbohydrates have unique roles to play within living systems and actively control a whole range of biological processes including cell growth and differentiation. The special biological properties of carbohydrates can be partially attributed to the location of the carbohydrates within biological systems—carbohydrates coat the majority of cell-surface and secreted proteins, so the first interaction a cell has with its environment will involve carbohydrates. A range of carbohydrate binding proteins, called lectins, exist, and these lectins mediate the interactions of cells with their environment, primarily via their initial interaction with the carbohydrates on the surfaces of cells. These interactive processes therefore play fundamental roles in cell–cell recognition and interaction processes.

Intriguingly, each type of cell displays **different** carbohydrates at its cell-surface and therefore cell surface carbohydrates act as distinguishing markers for each cell type. When a cell becomes diseased, a change in cell-surface carbohydrate structure can almost always be observed, raising the question *'Are carbohydrates directly responsible for the unusual properties of some diseased cell types?'* In Chapter 2 we will look in detail at the effect of carbohydrate structure on the properties of specific diseased states and see that this is indeed often the case.

Therapeutic potential

As well as providing solutions to some unanswered scientific questions, a study of carbohydrates may even offer novel opportunities for therapeutic treatments of diseases which are at present difficult to treat. For example, strategies are being developed which offer renewed hope for the treatment of diseases including cancer, AIDS and diabetes, and these strategies involve carbohydrates. Our progress in developing such therapies, however, relies upon rapid and efficient synthesis of oligosaccharides, necessitating the development of efficient synthetic or enzymatic methods. This is not a straightforward problem, but in this book we hope to provide an overview of recent methods and strategies which may make the syntheses of large carbohydrates realistic targets. Carbohydrate chemistry and biochemistry is a rapidly developing area and there is plenty of room for further contributions within this field. After all, without large quantities of biologically important oligosaccharides, a full understanding of their biological properties will be very difficult to assess.

Before we embark upon a study of carbohydrate chemistry, there are a few terms of nomenclature which need definition. The chemical and biochemical literature can appear daunting and filled with scientific jargon which is incomprehensible to the newcomer.

Nomenclature

In the same way that the terms '*amino acids*', '*peptides*' and '*proteins*' are used to describe similar compounds of different sizes, carbohydrates can be subdivided into '*monosaccharides*', '*oligosaccharides*' and '*polysaccharides*'. The monosaccharides represent the simplest monomer building blocks which are assembled to form larger structures, and a simple example with which you may be familiar is glucose. A range of monosaccharides occur naturally and serve as important building blocks for the synthesis of large oligosaccharides. Oligosaccharides are formed when 2–10 monosaccharides are joined together, for example, sucrose is a naturally occurring disaccharide, whilst polysaccharide is the term given to larger structures. Cellulose and glycogen are two polysaccharides which are naturally occurring. Their structures are so huge that we will not attempt to represent them diagrammatically!

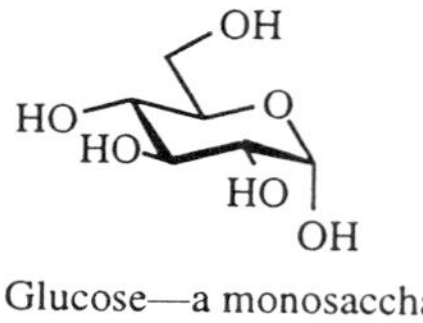

Glucose—a monosaccharide

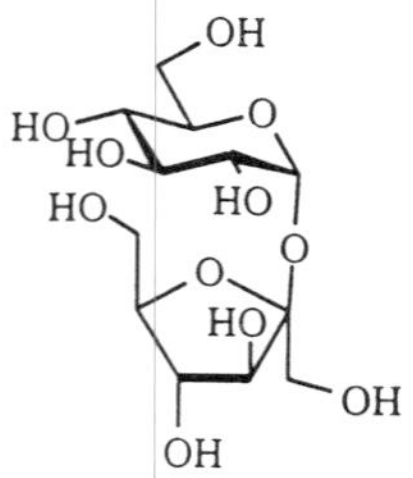

Sucrose—a disaccharide

Figure 1.1

Structural representation

You may well be puzzled by the range of different representations which are used to portray carbohydrate structures in the literature! As carbohydrate chemistry has developed, different structural representations have moved in and out of fashion so that there are many ways to represent identical molecules. Each different structural representation has its own merits, but throughout this book we will draw pyranose sugar units in their chair conformations whenever possible. For reference, different structural representations of one specific monosaccharide, glucose, are portrayed below in Figures 1.2–1.5.

When glucose is drawn as a chair conformation it is easy to depict the orientation of the bonds at the C-1 anomeric position. This is particularly important when we describe biologically important oligosaccharides. Bonds lying on the *opposite* side of the molecule to the C–H bond at the

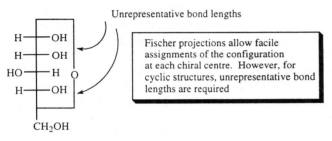

Fischer projection

Figure 1.2

Unrepresentative bond lengths

Fischer projections allow facile assignments of the configuration at each chiral centre. However, for cyclic structures, unrepresentative bond lengths are required

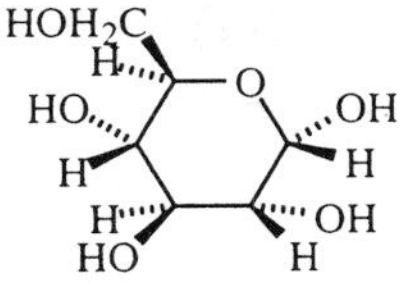

Wedge – slash formulae allow more representative illustration of bond lengths. However, no indication of the actual conformation of the pyranose ring is given. Structures are often cluttered.

Wedge – slash formula

Figure 1.3

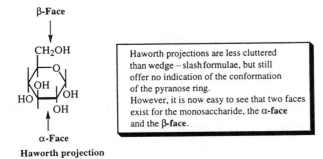

β-Face

α-Face

Haworth projection

Haworth projections are less cluttered than wedge – slash formulae, but still offer no indication of the conformation of the pyranose ring.
However, it is now easy to see that two faces exist for the monosaccharide, the α-**face** and the β-**face**.

Figure 1.4

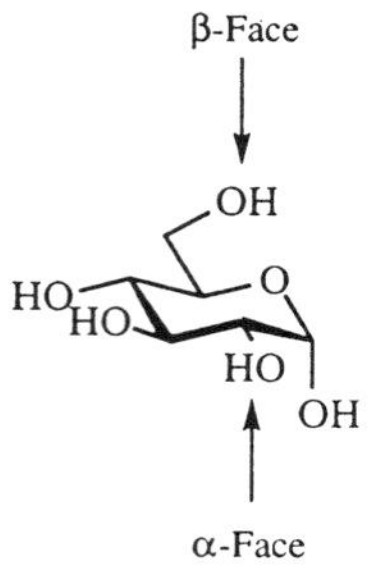

β-Face

Chair conformation

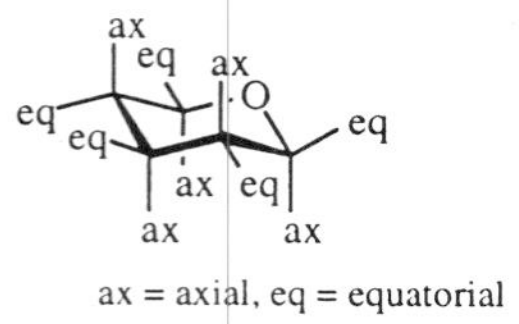

ax = axial, eq = equatorial

Chair conformations allow representation of the pyranose conformation. Hydroxyl substituents are placed in axial or equatorial positions. It is again easy to see that two faces exist, the **α-face** and the **β-face**.

α-Face

Figure 1.5

highest numbered asymmetric carbon atom (in this case, C-5) exist in the β-configuration, whilst bonds lying on the *same* side of the molecule as the C–H bond at the highest numbered asymmetric carbon atom exist in the α-orientation. This is discussed in further detail in Chapter 5.

Shorthand notations which are useful for representing large oligosaccharide structures are also often encountered in the literature. These abbreviations allow us to describe which monosaccharide units are linked together, to state the positions at which the linkages occur, and to define the geometry of the newly formed glycosidic bonds. For example, consider the disaccharides highlighted below.

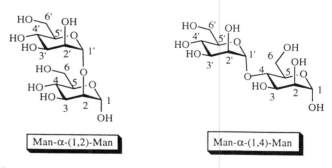

Man-α-(1,2)-Man

Man-α-(1,4)-Man

Figure 1.6

We could represent these structures in shorthand form as Man-α-(1, 2)-Man and Man-α-(1, 4)-Man respectively. Thus the Man-α-(1, 2)-Man disaccharide contains a mannopyranoside unit linked *via* the hydroxyl group at C-2 to C-1 of a further mannopyranoside unit, with the newly formed glycosidic bond existing in the α-configuration.

Problems associated with oligosaccharide assembly

A range of monosaccharide building blocks are present in nature and are therefore commercially available in their pure form. Naturally, many syntheses of oligosaccharides therefore involve the joining together of these monosaccharide units to furnish larger structures. At first sight this sounds like a very trivial process which must surely have been solved many years ago! Evidently, a number of complicating factors must be associated with this seemingly simple procedure which basically requires the reaction of one hydroxyl group of a monosaccharide acceptor with the anomeric centre of a further monosaccharide donor to furnish a disaccharide unit.

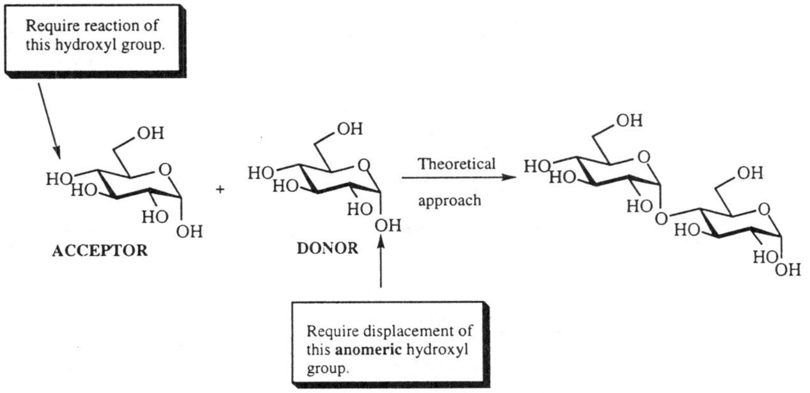

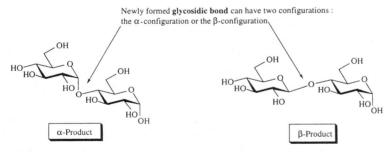

Figure 1.7

First, there are many different hydroxyl groups of similar reactivity on the monosaccharide acceptor unit which can all react at the anomeric centre of the donor to form a range of oligosaccharides. The acceptor must therefore be carefully controlled so that only the required hydroxyl group reacts. Even then, two possible diastereomeric disaccharides can result, corresponding to the α- and β-anomers of the product.

Figure 1.8

Thus the spatial orientation of the newly formed bond must also be carefully controlled to allow access to the specific isomer of choice.

Thirdly, each monosaccharide unit can exist in a number of structural forms, all of which are interconvertable. The pyranose form (six-membered ring) is in equilibrium with its isomeric furanose form (five-membered ring), as well as with the open chain aldehyde or ketone form (Figure 1.9).

Carbohydrate synthesis should never, therefore, be considered a simple problem! It is complications such as those outlined above which initially hampered progress in oligosaccharide synthesis. However, as the biological importance of oligosaccharides became better appreciated, it was

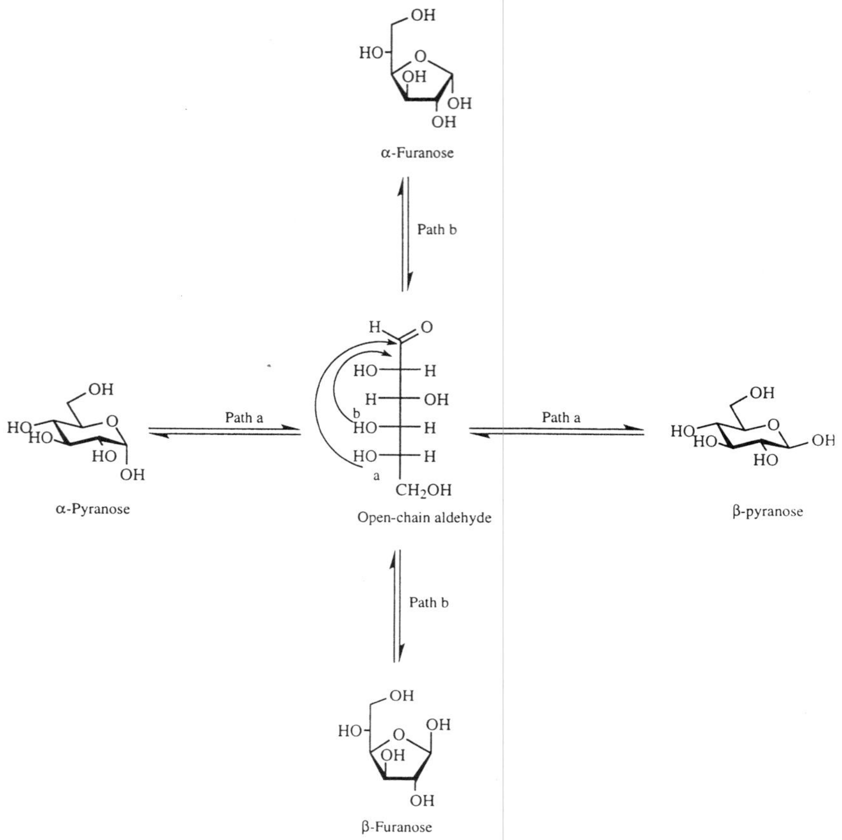

Figure 1.9

clear that chemists should investigate this highly challenging area of
chemistry and tackle the problems associated with carbohydrate synthesis. It is thanks to the dedication of many research workers that an
armoury of reactions are now available for modern carbohydrate
chemists to use in carbohydrate synthesis. However, the development of
new methods is still necessary, as shown by the continued identification
of new biologically important carbohydrates of even larger size.
Throughout this book we will show you some of the ingenious procedures which have been developed to solve the problems presented by
oligosaccharide synthesis. We hope that these will inspire you to develop
your own strategies and apply them to the synthesis of oligosaccharides of
direct relevance to your own research programmes.

Further reading

1. McNaught, A.D. (1997). Nomenclature of carbohydrates. *Advances in
   Carbohydrate Chemistry and Biochemistry*, **52**, 47–178.

2. Collins, P.M., Ferrier, R.J. (1995). *Monosaccharides*. John Wiley and
   Sons, Chichester.

3. Boons, G-J. (1998). *Carbohydrate chemistry*. Blackie Academic and
   Professional, London.

2 Biological roles of oligosaccharides

Oligosaccharides are fascinating biomolecules which are gaining in importance at a phenomenal pace compared with other well-studied biomolecules such as proteins and oligonucleotides. The current interest in oligosaccharides is a consequence of recent developments in sensitive and informative analytical techniques such as matrix-assisted laser-desorption ionisation–time of flight MALDI–TOF spectroscopy, 3D-nuclear magnetic resonance (n.m.r.), and capillary electrophoresis. Furthermore, this interest is driven by the constant need to further understand the biological roles of oligosaccharides, since they are reported to be involved in many fundamental processes including immune defence, fertilisation, viral replication, parasite infection, cell growth, cell–cell adhesion, degradation of blood clots and inflammation. One major obstacle in the study of biologically important oligosaccharides is the low availability of stereochemically pure samples from natural sources. However, with the exciting development of synthetic strategies which allow access to comparatively large quantities of natural and unnatural oligosaccharides, further advances in glycobiology are anticipated in the future.

It would be an immense task to tackle all the topics pertaining to oligosaccharides, such as structural analysis, protein conformation stabilisation, occurrence in plants and animals, and their metabolism, in one chapter of this book. Thus, the aim of this chapter is to describe general, rather than specific, properties of oligosaccharides. Some human diseases which are believed to involve oligosaccharides in their progression will also be described, to show how carbohydrate chemists can provide novel ways for treating these diseases.

Biosynthesis of some biologically important oligosaccharides

The biosynthesis of oligosaccharides is governed by a whole range of enzymes such as glycosyl transferase, oligosaccharyl transferase and glycosidases. Oligosaccharide biosynthesis commences by employing a series of glycosyl transferase enzymes to join saccharide units together in regio- and stereoselective reactions (Figure 2.1). These enzymes work to assemble a $Glc_3Man_9GlcNAc_2$ oligosaccharide which is covalently bound to a lipid. This is subsequently transferred from the lipid to a peptide, via an oligosaccharyl transferase enzyme. The $Glc_3Man_9GlcNAc_2$ oligosaccharide is further processed by a series of glycosyl hydrolase and glycosyl transferase enzymes to afford a wide range of structurally diverse oligosaccharides. This occurs during transport of the glycoprotein through the endoplasmic reticulum and Golgi apparatus. The structure of the

Glc$_3$Man$_9$GlcNAc$_2$–Lipid

Oligosaccharyl transferase

Glc$_3$Man$_9$GlcNAc$_2$–Peptide

α-Glucosidase I and II

Man$_9$GlcNAc$_2$–Peptide

α-Mannosidase

The biosynthesis of oligosaccharides involves glycosyl transferase, oligosaccharyl transferase and glycosidase enzymes.

High-mannose-type oligosaccharides ← Man$_8$GlcNAc$_2$–Peptide

α-Mannosidase I and GlcNAc transferase I

Hybrid-type oligosaccharides ← GlcNAcMan$_5$GlcNAc$_2$–Peptide

α-Mannosidase II

Complex-type oligosaccharides ← GlcNAcMan$_3$GlcNAc$_2$–Peptide

Figure 2.1

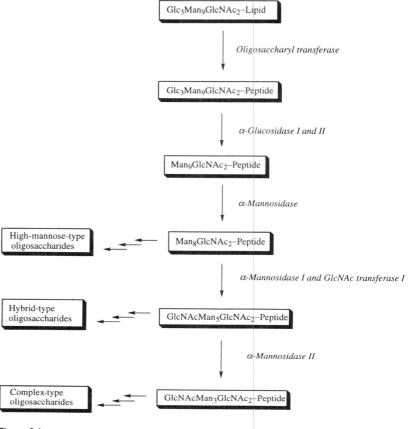

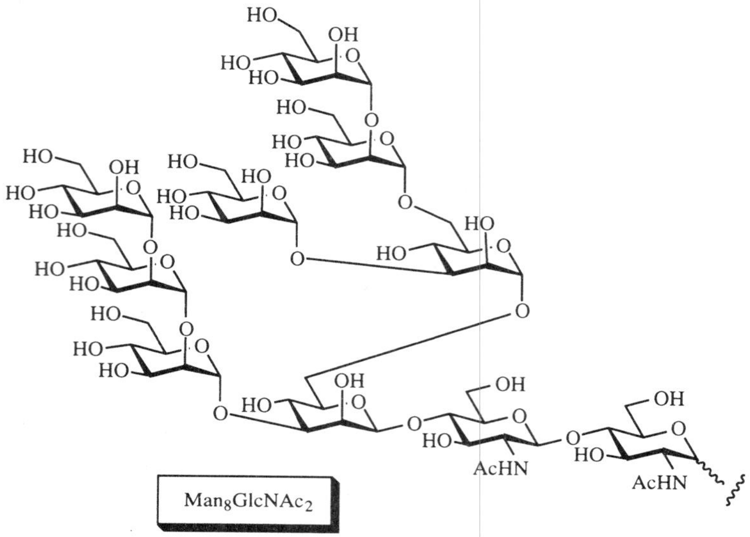

Man$_8$GlcNAc$_2$

Figure 2.2

Man$_8$GlcNAc$_2$ oligosaccharide, which is a precursor to high mannose, hybrid and complex-type oligosaccharides, is illustrated in Figure 2.2.

Since the structures of the oligosaccharides and conjugates are determined by a series of enzymes, they are also delicately controlled by the genes expressing the particular glycosyl transferase and glycosidase enzymes. It appears that the delicate balance of enzymes causing biosynthesis of the oligosaccharides can be easily disrupted in a diseased state, giving rise to a changed expression of oligosaccharide structure on the cell surface. For example, tumour cells often express high levels of sialic acid on their surface. Consequently, oligosaccharide expression may be used as a marker for the diseased state, and this will be considered in more detail later.

A variety of monosaccharides can be incorporated into oligosaccharides via glycosyl transferases, and some of the more common monomer units available are shown in Figure 2.3.

The newly formed oligosaccharides can undergo further enzymic modification to allow introduction of sulfate, phosphate or acetate groups. For example, the proteoglycan heparin is a highly sulfated oligosaccharide responsible for blood clotting (Figure 2.4).

Glycoconjugates

In the body, oligosaccharides are either dissolved in the aqueous media surrounding all cells, as free macromolecules, or are tethered to other biomolecules to afford glycoconjugates. The exact role of the carbohydrate portions of glycoconjugates is still poorly understood. Nevertheless, they have been implicated in cell adhesion, growth control and regulation of receptors. Covalent linkage of oligosaccharides to lipids and proteins affords glycolipids and glycopeptides, respectively. The oligosaccharides impart special properties to the biomolecules to which they are bound. For example, once bound as glycolipids, lipids become more water soluble and can behave like detergents, readily forming micelles at high concentrations. As a result, glycolipids are often found embedded in the bi-layer of cell membranes, with the oligosaccharide protruding into the aqueous solution on the outside. Glycolipids often contain sphingosine as the lipid, and this is capable of embedding in the cell membrane (Figure 2.5). Glycosphingolipids are believed to play key roles in cell–cell adhesion, cell–cell recognition, antigenic specificity and other types of transmembrane signalling.

Glycopeptides can be divided into two main categories, depending on whether the saccharide is *N*-linked to an amino acid, such as L-asparagine, or *O*-linked to an amino acid such as L-threonine or L-serine (Figure 2.6).

Proteoglycans are a subclass of glycoproteins which are polyanionic, are of relatively large size and display repeating units. Heparin, illustrated previously in Figure 2.4, is one example of a proteoglycan involved in blood clotting. The cell wall structure of bacterial cells is another important example of proteoglycans.

Carbohydrates as carriers of biological information

Due to their high water solubility, oligosaccharides are located on the outer surface of cell membranes. It is therefore logical that carbohydrates may be involved in cell–cell recognition and cell differentiation. Moreover, there

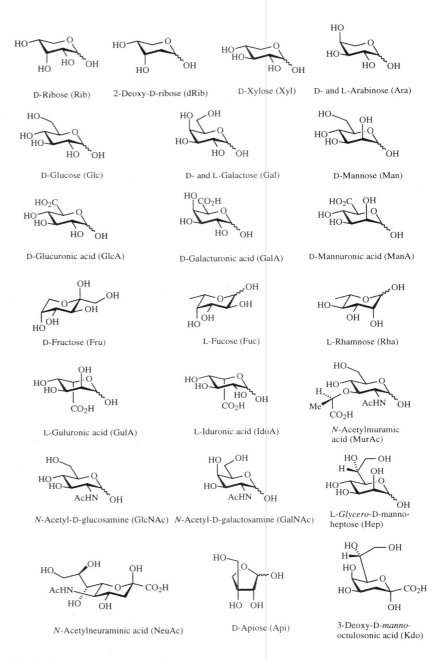

D-Ribose (Rib)   2-Deoxy-D-ribose (dRib)   D-Xylose (Xyl)   D- and L-Arabinose (Ara)

D-Glucose (Glc)   D- and L-Galactose (Gal)   D-Mannose (Man)

D-Glucuronic acid (GlcA)   D-Galacturonic acid (GalA)   D-Mannuronic acid (ManA)

D-Fructose (Fru)   L-Fucose (Fuc)   L-Rhamnose (Rha)

L-Guluronic acid (GulA)   L-Iduronic acid (IdoA)   *N*-Acetylmuramic acid (MurAc)

N-Acetyl-D-glucosamine (GlcNAc)   *N*-Acetyl-D-galactosamine (GalNAc)   L-*Glycero*-D-manno-heptose (Hep)

N-Acetylneuraminic acid (NeuAc)   D-Apiose (Api)   3-Deoxy-D-*manno*-octulosonic acid (Kdo)

Figure 2.3

are a huge number of ways in which monosaccharides can combine to afford oligosaccharides. Unlike other biomolecules which appear to have ordered structures (e.g. the G–C and A–T base pair combinations of DNA or the tri-base gene sequence rule for particular amino acids), no such rules have been identified which control the formation of oligosaccharide structures. However, nature has not been completely random in its selection of monosaccharides and their use in building complex

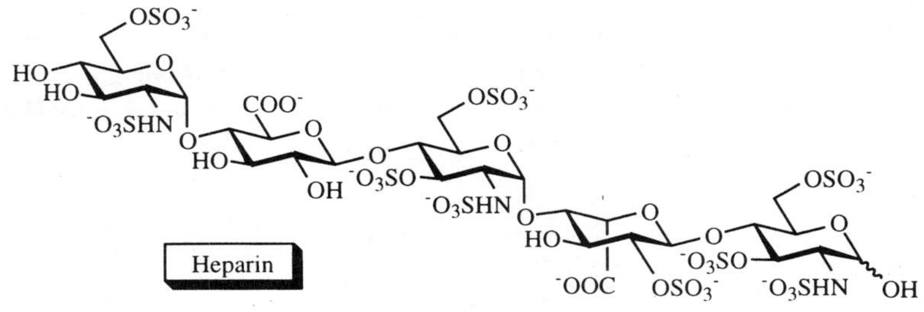

Figure 2.4

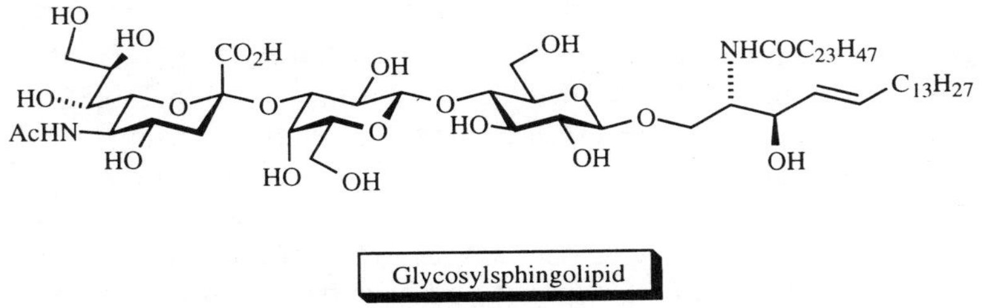

Glycosylsphingolipid

Figure 2.5

Carbohydrates can combine with lipids and peptides to form glycolipids and glycopeptides, respectively.

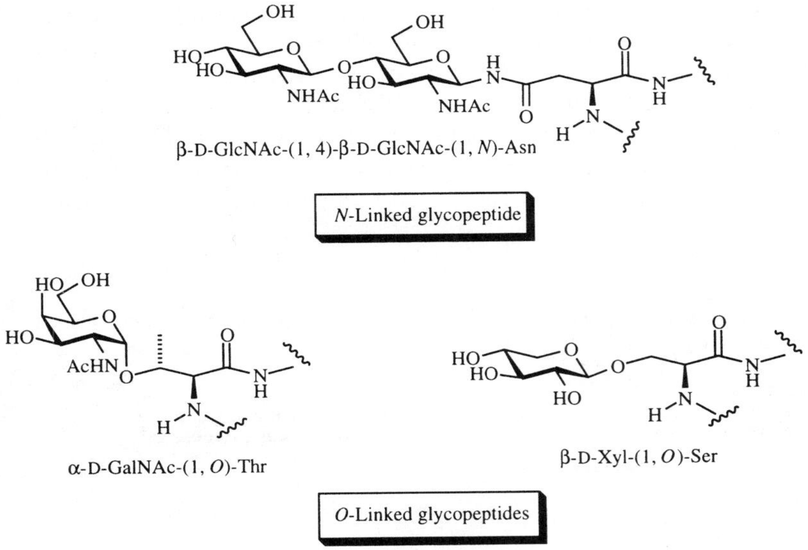

β-D-GlcNAc-(1, 4)-β-D-GlcNAc-(1, *N*)-Asn

N-Linked glycopeptide

α-D-GalNAc-(1, *O*)-Thr

β-D-Xyl-(1, *O*)-Ser

O-Linked glycopeptides

Figure 2.6

oligosaccharides. First, D-monosaccharides are usually incorporated within the oligosaccharides rather than the enantiomeric L-series, as chosen for amino acids in natural proteins. Secondly, the monosaccharides tend to exist in the pyranose (six-membered ring) or furanose (five-membered ring) forms, rather than smaller ring systems. Thirdly, monosaccharides are always linked via the anomeric position to form glycosidic or acetal bonds. Even within these constraints, there are far more ways to

Table 2.1

Combination of individual units	Number of saccharides possible	Number of peptides possible
Two identical units A–A dimer	11	1
Three identical units A–A–A trimer	176	1
Three different units A–B–C trimer	1056	6

link monosaccharides together to form oligosaccharides than offered when amino acids join together to form proteins as illustrated in Table 2.1.

Thus oligosaccharides are excellent molecules for transporting biological messages between a variety of cells and also within cellular compartments.

Diverse structure alone is not sufficient to allow good representation of biological information. A distinct spatial organisation and considerable structural rigidity are also required and have been shown to be present within oligosaccharides and their conjugates. There is evidence in nature to confirm that carbohydrates are indeed involved in dynamic biological messages and thus lead to cell differentiation and cell–cell recognition processes. Moreover, it has been discovered that oligosaccharides of even short sequences are used for carrying important biological information. For example, human blood groups are differentiated by relatively simple changes in oligosaccharide structure as illustrated in Figure 2.7.

Furthermore, the host's immune system uses carbohydrate markers to distinguish its own cells from foreign cells, highlighting the importance of matching blood groups during blood transfusion and organ transplants.

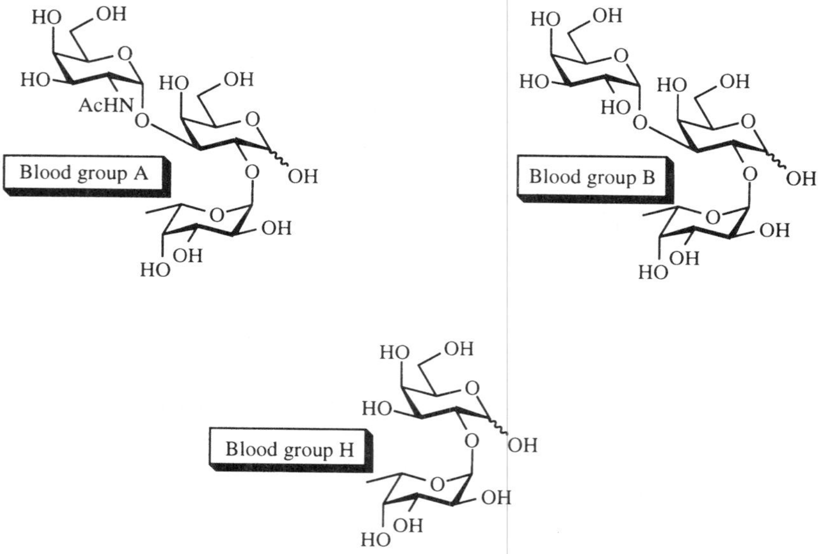

Carbohydrates are involved in cell differentiation and cell–cell recognition processes.

Figure 2.7

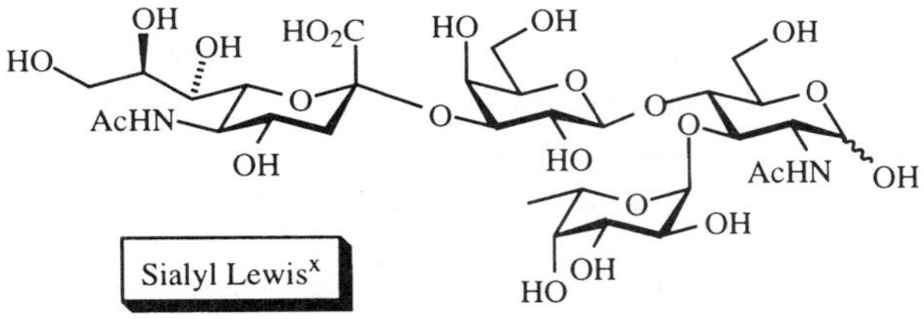

Figure 2.8

The structure of the tetrasaccharide expressed on the surface of white blood cells, sialyl Lewis^x, is depicted above.

Further evidence confirming that oligosaccharides are involved in cell differentiation can be provided by studying the development of a foetus in the womb. During this process, large quantities of oligosaccharides can be detected in the surrounding placenta. The structure and occurrence of these oligosaccharides varies as the development of the foetus progresses. It is now believed that this is a result of carbohydrates being used both to differentiate cells which are at different stages of development and to differentiate cells programmed to form different organs.

Although the variation in oligosaccharide structure provides a powerful tool for carrying biological information, it is not sufficient, alone, to completely account for the many biological roles of carbohydrates. The oligosaccharides on the surface of cells must have some mechanism which allows them to interact with other cells. For example, viruses, bacteria, toxins and parasites, which often display carbohydrates on their surfaces, must be able to interact with their hosts to initiate infection. Therefore receptors for carbohydrates must be present within living systems and must play fundamental roles in recognition processes. These receptors are called lectins.

Lectins: carbohydrate-binding proteins

Initially, carbohydrates were only considered of importance by the food industry where they were used as high-energy, nutritional foods. However, the isolation of a plant protein from *Canavalia ensiformis* extracts in 1936, which bound to carbohydrates on erythrocytes, red blood cells, and caused them to precipitate, raised the interest of glycobiologists, who then initiated a search for plant proteins capable of binding alternative sugars. It was soon demonstrated that different plant proteins had affinities for specific carbohydrates present on the surface of erythrocytes from different individuals, hence corresponding to an individuals' blood group. These carbohydrate-binding proteins were termed 'lectins'. This carbohydrate-binding property of the lectins was considered a special property of only plants, and was a useful means of identifying the blood group of individuals. However, in 1982 the first animal lectins were discovered and this breakthrough stimulated the present phenomenal activity in the oligosaccharide area of research. Lectins are found on the surface of all cell membranes, for example liver cells, macrophages, and epithelial and tumour cells, and can interact with oligosaccharides present on surrounding cells. Hence they

Lectins are carbohydrate-binding proteins.

play key roles in molecular recognition. The density of lectins on the surface of cells varies and, if low, may be a limiting factor in their identification and isolation from human tissue samples. Certain biological effects, particularly in multicellular organisms, are the result of a particular oligosaccharide ligand docking into its complimentary lectin/receptor on a neighbouring cell.

The plant and animal lectins are composed of amino acids and contain very little homology, but have been found to interact with the oligosaccharides via Hydrogen bonding, metal coordination, van der Waals forces and hydrophobic interactions. The availability of large numbers of hydroxyl groups on sugars allows favourable hydrogen bonding with the protein's amino acids, particularly with Asp and Asn.

If oligosaccharide:lectin binding is to prove effective for molecular recognition, specific rather than random binding must occur. Since lectins bind only weakly with monosaccharides (in the mM range), selectivity must be a result of more subtle factors. One popular view is that selectivity is increased through multiple binding, for instance through additional binding in subsites. In this case, the terminal monosaccharide is bound by the primary site of the lectin but additional monosaccharides along the carbohydrate chain bind to subsites on the lectin. Thus although the terminal monosaccharide causes primary recognition, the overall strength of the binding is dependent upon the composition of the oligosaccharide. In this way lectins which occur on the surface of cell membranes as differentiation markers can bind selectively to particular oligosaccharides, and initiate a biological response.

Lectins are involved in specific, selective binding with carbohydrates rather than random binding.

Examples of carbohydrate:lectin interactions

Oligosaccharide:lectin binding plays a fundamental role in the conception of human life. Union of the egg and sperm is of low probability, but the chances of fertilization are improved by chemoattractants produced by the egg which attract the sperm. Oligosaccharides present on the surface of the sperm cells allow the sperm to interact with lectins on the egg's surface and initiate fertilisation of the egg. These oligosaccharides interact by specific recognition of α-galactosyl and α-fucosyl end-groups by the spermatozoa receptors. It is interesting that once the egg has been fertilised, it then releases an enzyme which cleaves the oligosaccharides from the surface of the sperm cells so that further fertilisation is discouraged.

Such oligosaccharide:lectin interactions are essential for smooth and efficient biological processes such as embryo development and removal of damaged cells. It is interesting that pathogenic bacteria, parasites and viruses have evolved mechanisms which use this important interaction for their own survival and propagation. For example, the oligosaccharides found on the surface of bacteria, parasites or viruses can interact with lectins on host cells to allow infective processes to occur. It is pertinent to see that nature provides a mechanism for aborting this infective process in breast-fed babies. Human milk contains many soluble oligosaccharides and breast fed babies are exposed to high levels of these oligosaccharides, especially in their first few days. The composition of these oligosaccharides is very complex, and their role is presumably more than just nutritional value. It is believed that these molecules bind to a wide range of lectins on the

Some infective processes also involve carbohydrate–lectin binding.

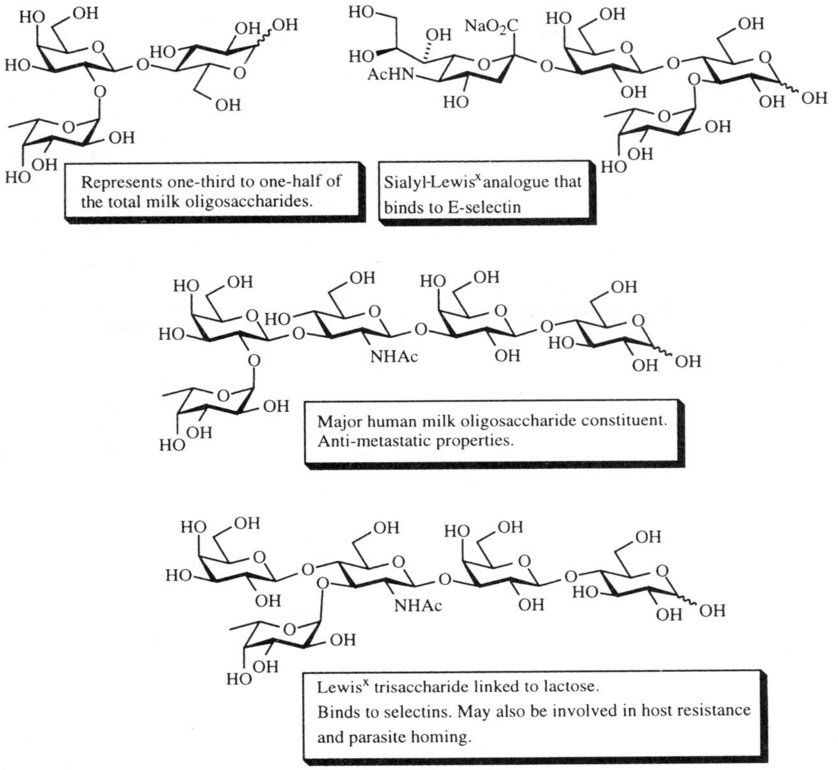

Figure 2.9

surface of the epithelial cells lining the mouth, oesophagus and stomach, and throughout the gastrointestinal system in the new born baby.

This in turn prevents opportunistic infections whilst the baby's immune system is developing. Degradation of the foreign species by the liver can then occur without any infection occurring, so long as the potentially infective species remain in the blood and are unable to invade other organ cells. It may also be that these soluble oligosaccharides are priming the baby's immune system to tolerate the oligosaccharides expressed on the surface of the baby's differentiated cells in the various organs.

The soluble milk oligosaccharides which inhibit infection have been termed 'anti-infective' or 'anti-adhesive' oligosaccharides. Many scientists are now trying to mimic this anti-infective process and produce large quantities of oligosaccharides capable of interfering with other infective processes. For example, bacterial infections caused by MS fimbriae such as *E. coli* 0157 H7 and certain Salmonella strains recognise oligosaccharide receptors *in vivo* that contain mannose units. Hence anti-infective oligosaccharides containing mannose could potentially inhibit the infective process (as highlighted in Figure 2.10).

Oligosaccharide–lectin binding can also be used to target therapeutic agents to diseased cells which express high densities of specific lectins on their surface. For example, GalNAc-clusters have been used to target antisense-nucleotides to hepatocytes to potentially allow treatment of hepatitis A. The trivalent clusters illustrated in Figure 2.11 have been utilised in this

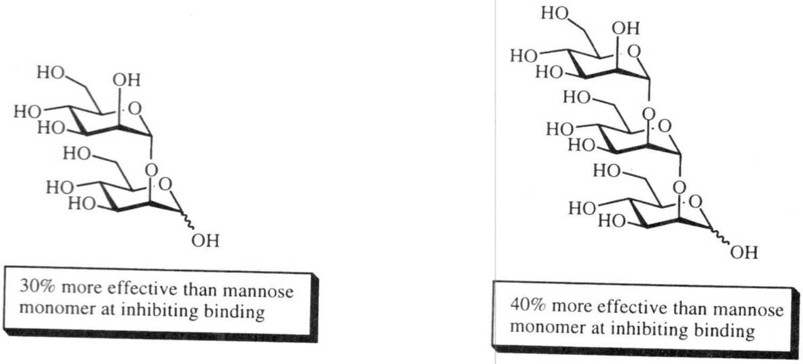

30% more effective than mannose monomer at inhibiting binding

40% more effective than mannose monomer at inhibiting binding

Figure 2.10

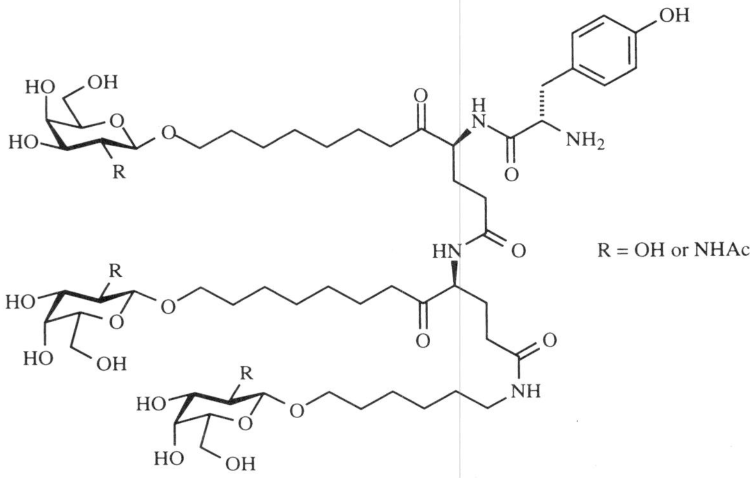

Carbohydrate–lectin binding can be used to target therapeutic agents to some diseased cells.

R = OH or NHAc

Figure 2.11

approach because they are able to bind to the required lectins as strongly as galactose-terminated multiantennary oligosaccharides. After conjugation to therapeutic agents, the affinity of the lectins for the small oligosaccharide fragments allows targeting of the therapeutic agents to the lectins, and hence to the diseased state.

Diseases in which oligosaccharides play an important role

As mentioned above, the density of cell-surface oligosaccharides is often altered in diseased states. Some examples are provided below which illustrate the effect of varying oligosaccharides on the surface of diseased cells. Access to oligosaccharides displayed on the cell surface can be invaluable to glycobiologists for a number of reasons. First, it may allow full investigation of their biological roles *in vitro*. If access to novel oligosaccharide analogues and derivatives is possible, this may also allow manipulation of the diseased state to improve the human condition. Even more exciting, synthesised oligosaccharides may allow the development of vaccines to fight the diseases with which they are associated. We will consider some of these theories in the concluding sections of this chapter.

Acquired Immunodeficiency Syndrome (AIDS)

A glycoprotein with a molecular weight of 120 kilodaltons (gp120) has been found to play an important role in the infectivity of HIV, one of the infective viral strains causing AIDS. It has been demonstrated that this glycoprotein is involved in the initial binding of the virus to lectins on the surface of circulating T-lymphocytes (T-cells), particularly the CD4 subpopulation, which allows subsequent cell invasion. Once inside the T-cell, the virus replicates safely with no possibility of attack by the host's immune system. The mature virus then causes the T-cell to burst open, releasing thousands of virus particles into the blood stream, ready to infect other T-cells by the same gp120/CD4 mechanism. When this is repeated over many years, the overall result is disintegration of the patient's immune system, thus eventually making them vulnerable to opportunistic infections and other diseases. Lectins were widely used to characterise the oligosaccharide structures present on gp120 and also to identify those involved in CD4-gp120 binding. As a result it was found that the glycoproteins are composed of four major classes of oligosaccharides, namely a high mannose-type glycan, a bisected hybrid-type glycan, a biantennary fucosylated complex-type glycan and a triantennary bisected complex-type glycan (as illustrated in Figure 2.12).

Gp120 is involved in the binding of the AIDS virus to lectins on T-cells.

Drugs currently available for treatment of AIDS, such as AZT, target the crucial viral DNA synthetic step during the replication of the AIDS virus. The enzymes associated with DNA synthesis incorporate the AZT molecules rather than 'real' nucleic acid building blocks so that synthesis of DNA associated with the AIDS virus is aborted (Figure 2.13).

Severe side effects are encountered during this treatment regime, which is expensive since the drug has to be continuously administered until all the viral particles have been destroyed, and of limited efficacy, since development of resistant viral strains often ensues. A more hopeful and economical approach for the treatment of HIV infection may be to generate vaccines. The antigens on the virus coat have been studied for this purpose and gp120 has been found to be involved in the important initial binding step. It is both an oligosaccharide and a protein, thus both moieties are potential vaccination antigens. The oligosaccharide has been found to be mainly *N*-linked to an asparagine residue in the protein, and isolation of oligosaccharides from many strains of HIV particles has shown that there are a number of oligosaccharides present. However, the number of glycosylation sites on the protein component of gp120 is largely conserved. Generation of antibodies to both the protein and the carbohydrate components of gp120 is underway by many groups and it is hoped that an antigen which is both unique to HIV and imparts a strong immune response can be discovered. The oligosaccharides are excellent targets for a vaccination program since they are located on the outside of the HIV envelope glycoprotein, and are thus readily accessible for recognition by the primed immune system. The glycoprotein gp120 has a number of immunogenic sites and an ideal vaccination antigen would be a conserved region amongst the various strains of HIV.

Synthetic access to oligosaccharides present on the AIDS virus may allow development of a vaccine for AIDS.

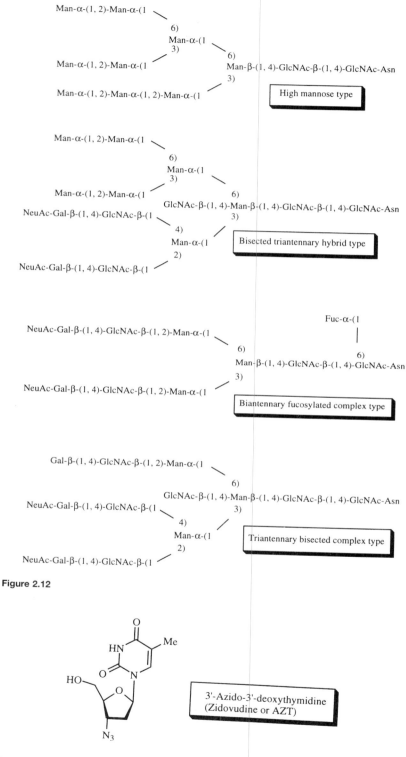

Man-α-(1, 2)-Man-α-(1
　　　　　　　　　6)
　　　　　　Man-α-(1
　　　　　　　　　3)
　　　　　　　　　　　6)
Man-α-(1, 2)-Man-α-(1
　　　　　　　　　　　Man-β-(1, 4)-GlcNAc-β-(1, 4)-GlcNAc-Asn
　　　　　　　　　　　3)
Man-α-(1, 2)-Man-α-(1, 2)-Man-α-(1

High mannose type

Man-α-(1, 2)-Man-α-(1
　　　　　　　　　6)
　　　　　　Man-α-(1
　　　　　　　　　3)
　　　　　　　　　　　6)
Man-α-(1, 2)-Man-α-(1
GlcNAc-β-(1, 4)-Man-β-(1, 4)-GlcNAc-β-(1, 4)-GlcNAc-Asn
NeuAc-Gal-β-(1, 4)-GlcNAc-β-(1
　　　　　　　　　　　3)
　　　　　　　　　4)
　　　　　　Man-α-(1
　　　　　　　　　2)

Bisected triantennary hybrid type

NeuAc-Gal-β-(1, 4)-GlcNAc-β-(1

　　　　　　　　　　　　　　　Fuc-α-(1
NeuAc-Gal-β-(1, 4)-GlcNAc-β-(1, 2)-Man-α-(1
　　　　　　　　　　　　　　　　　|
　　　　　　　　　　　6)
　　　　　　　　　　　　　　　　　6)
　　　　　　　　　Man-β-(1, 4)-GlcNAc-β-(1, 4)-GlcNAc-Asn
　　　　　　　　　　　3)
NeuAc-Gal-β-(1, 4)-GlcNAc-β-(1, 2)-Man-α-(1

Biantennary fucosylated complex type

Gal-β-(1, 4)-GlcNAc-β-(1, 2)-Man-α-(1
　　　　　　　　　　　6)
　　　　　　　　　GlcNAc-β-(1, 4)-Man-β-(1, 4)-GlcNAc-β-(1, 4)-GlcNAc-Asn
NeuAc-Gal-β-(1, 4)-GlcNAc-β-(1
　　　　　　　　　　　3)
　　　　　　　　　4)
　　　　　　Man-α-(1
　　　　　　　　　2)

Triantennary bisected complex type

NeuAc-Gal-β-(1, 4)-GlcNAc-β-(1

Figure 2.12

3'-Azido-3'-deoxythymidine
(Zidovudine or AZT)

Figure 2.13

Cancer

A number of tumour-associated antigens have been identified in recent years, and those that have been isolated and characterised are primarily glycoproteins. The most common antigens display the sialylated oligosaccharides (as illustrated in Figure 2.14 and Table 2.2).

The occurrence of these tumour-associated antigens is a result of the upregulation of particular genes within the cell. This consequently gives rise to an over-expression of particular glycosylation enzymes, thereby raising the expression of particular oligosaccharides such as sialylated oligosaccharides on the surface of the cells. Normally, sialylated antigens are found on activated T-cells and are only expressed during penetration of the lymphocytes to infected sites. Sialylated molecules can bind to selectins, which are expressed on the surface of epithelial cells lining the blood vessels (selectins are lectins which specifically bind particular oligosaccharides, for example E-selectin binds sialyl Lewis[x]). This allows the T-cells to stick to the blood-vessel walls and migrate, by a rolling motion, towards the site of damage or infection, and is therefore a beneficial process. However, as cancer cells express an abundance of sialylated oligosaccharides, they, too, can bind to lectins such as E-selectin on the surface of epithelial cells lining blood vessels. They can therefore also roll along the surface of the epithelial cells, which allows them to establish secondary tumours at a new site, distant from the primary tumour, in a

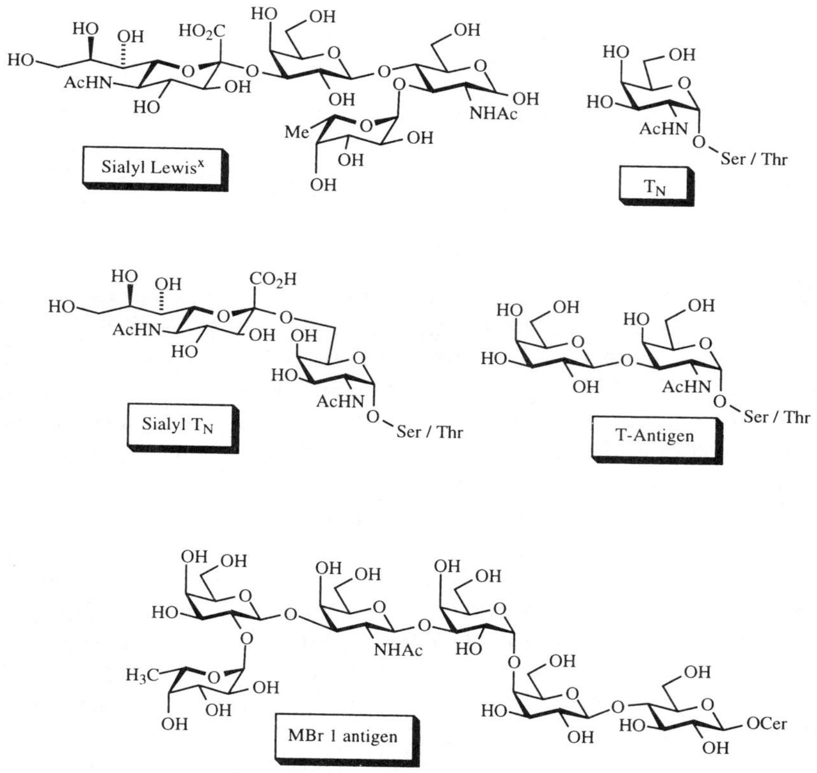

Figure 2.14

Table 2.2

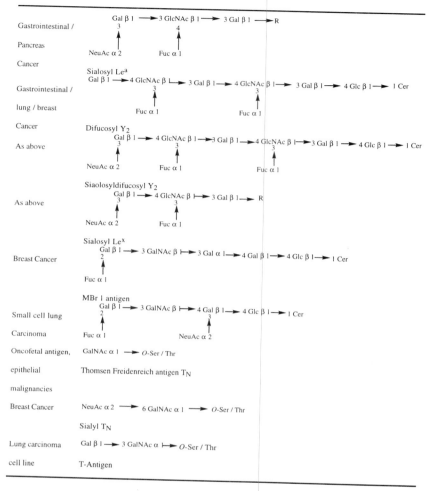

Gastrointestinal / Pancreas Cancer	Sialosyl Le^a
Gastrointestinal / lung / breast Cancer	Difucosyl Y$_2$
As above	Siaolosyldifucosyl Y$_2$
As above	Sialosyl Le^x
Breast Cancer	MBr 1 antigen
Small cell lung Carcinoma	
Oncofetal antigen, epithelial malignancies	GalNAc α 1 ⟶ *O*-Ser / Thr — Thomsen Freidenreich antigen T$_N$
Breast Cancer	NeuAc α 2 ⟶ 6 GalNAc α 1 ⟶ *O*-Ser / Thr — Sialyl T$_N$
Lung carcinoma cell line	Gal β 1 ⟶ 3 GalNAc α ⟶ *O*-Ser / Thr — T-Antigen

Cancer cells display high levels of sialylated oligosaccharides and these give the cancer cells unusual properties compared with normal cells.

detrimental process known as metastasis. Some novel cancer treatments attempt to prevent metastasis of the tumour by inhibiting binding to the surface of the epithelial cells in the blood vessels. In this approach, soluble oligosaccharides such as the tetrasaccharide sialyl Lewis^x are synthesised and administered to the patient. These block the E-selectin sites on the epithelial cells in the blood vessel so that the tumour cells are unable to spread from their primary site. Surgery can then be performed to remove the tumour. One drawback to this therapy is that sialyl Lewis^x is also expressed on normal cells and is used by many cells, such as T-lymphocytes, leucocytes and macrophages, to migrate into regions of injury and infection to allow repair or destruction of some of the naturally damaged cells. Consequently, if the soluble oligosaccharides are administered over a long period of time, some of the normal functions of the body may be seriously affected, resulting in chronic infection and gradual deterioration of vital organs in the body.

Sialyl Lewis^x oligosaccharides are also excellent markers for identifying metastatic tumours. These markers are very useful as quick diagnostic factors for identifying cancer patients and for monitoring their progress during cancer therapy. These glycoprotein markers are also very useful for the development of novel cancer therapies. The Carcinoma Embryonic Antigen (CEA) is another example of a tumour-associated antigen which is overexpressed in colon tumours. It is highly expressed during the development of the embryo but disappears at the later stages of childhood, and is only expressed in low levels in the gut of adults. However, colon tumours express high levels of this antigen, and antibodies to this antigen have been raised to target colon tumours. This has been investigated in a number of ways.

Targeted cancer therapies

One method of targeted therapy attaches toxins or cytotoxic agents to antibodies to selectively deliver them to the tumour sites. If the chosen antigen is located on the surface of the tumour cell, the therapeutic agent can, in theory, be delivered selectively to the tumour cells carrying these antigens. Unfortunately, there are some drawbacks to this approach. First, the tumour-associated antigens are also expressed, in low levels, on normal cells, leading to some non-localised toxicity. Secondly, as noted above, tumours have heterogeneous expression of surface markers so that incomplete targeting of the tumour may result. However, research is still continuing in this area.

A further use of tumour-associated antibodies involves conjugation of an enzyme, normally not found in the human body, to the tumour-associated antibody, to result in an antibody–enzyme conjugate. This conjugate is then administered into the patient and allowed to localise at the tumour site, and the excess allowed to clear from the circulation. Thus the antibody is used to direct the enzyme specifically to the tumour site. A non-toxic prodrug which is a substrate for the targeted enzyme is then administered. Enzyme-mediated release of the cytotoxic agent from the prodrug will occur only at the tumour site, allowing selective elimination of the tumour. This approach is called ADEPT and refers to Antibody-Directed Enzyme Prodrug Therapy. This targeting strategy has a superior advantage over the previously described antibody–drug-targeting approach and traditional chemotherapy. By generating high local concentrations of cytotoxic agent at the tumour site, even heterogeneous tumour cells are exposed to the cytotoxic agent. This approach is presently undergoing Phase I clinical trials by Zeneca Pharmaceutical and the outcome is being awaited with great excitement by clinical oncologists.

Diabetes

This is a serious disease in which the patient is unable to remove excess glucose from the blood stream. If this is not brought under control within a short time, damage of vital organs may result. After ingestion of carbohydrates, glycosidases such as sucrase and amylase degrade the oligosaccharides, allowing release of free glucose into the blood stream. Normally, the pancreas releases insulin to increase the conversion of glucose into glycogen. Genetic, or acquired, diabetes involves malfunction of this important process, and allows high levels of glucose to accumulate in the blood. The use of glycosidase inhibitors (see Figure 2.15),

Glycosidase inhibitors can be used to decrease the level of glucose in diabetics.

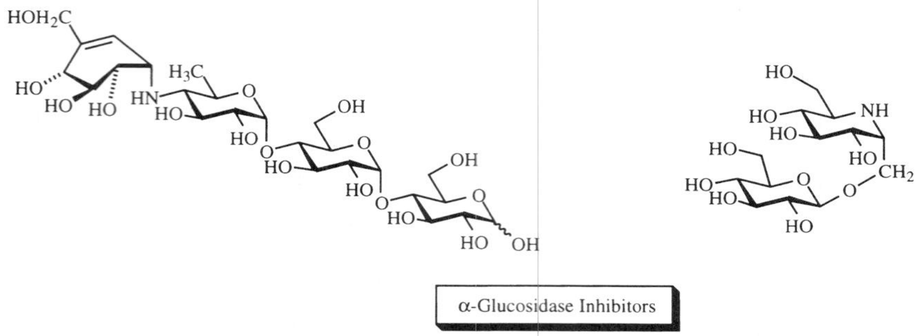

α-Glucosidase Inhibitors

Figure 2.15

which inhibit degradation of ingested carbohydrates, is one manner in which production of glucose can be minimised, and these inhibitors are being investigated in novel therapies for the treatment of diabetes.

The glycosidase inhibitors are, however, very potent molecules, which can cause severe side effects. This is because oligosaccharides are continuously being metabolised in the body, for a wide range of biological functions, and glycosidase inhibitors can potentially disrupt these processes ultimately causing a whole range of biological malfunctions.

Glycosidase inhibitors

We have already seen that a range of glycosyl transferase and glycosidase enzymes are involved in the biosynthesis of oligosaccharide-containing glycoproteins and glycolipids. We have also seen how these oligosaccharides impart unique properties on cells and are central to many biologically important processes. Inhibition of these enzymes has therefore received much attention in recent years. If the glycosyl transferase and glycosidase enzymes are inhibited, novel oligosaccharides on cell surfaces are produced, and the effect of these oligosaccharides on the properties of the cells can be examined. Furthermore, if the synthesis of oligosaccharides associated with specific diseases can be aborted, novel ways for treating the diseases associated with these oligosaccharides can be obtained. Such approaches have been studied for the treatment of AIDS, cancer and diabetes.

Aza sugars have received much attention for their inhibition of N-linked oligosaccharides. The structures of some potent inhibitors are highlighted below (see Figure 2.16).

It is believed that protonated aza sugars are transition-state analogues and can bind to the enzyme active site by charge–charge interactions and hydrogen bonding.

Glucosidase inhibitors such as castanospermine and 1-deoxynojirimycin have been used for modifying oligosaccharide on the gp120 glycoprotein which is associated with the AIDS virus. Such compounds efficiently inhibited the glucosidase trimming enzymes, which in turn allowed synthesis of unusual oligosaccharides with higher proportions of glucose on gp120. It was found that the interaction of gp120 with the lectin on the CD4 protein was altered and this consequently interfered with infectivity.

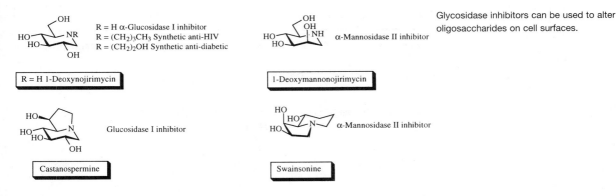

Glycosidase inhibitors can be used to alter oligosaccharides on cell surfaces.

Figure 2.16

As noted above, tumour cells often display high levels of sialylated oligosaccharides. The loss of such sialylated antennae and, consequently, synthesis of lectin-resistant phenotypes can be induced by treating tumour cells with swainsonine. Since swainsonine inhibits mannosidase II enzymes, unusual cell surface oligosaccharides with high levels of α-(1, 6)-Man and α-(1,3)-Man terminal residues rather than β-(1, 2)-sialyllactosamine and β-(1, 6)-sialyllactosamine terminal residues can be formed.

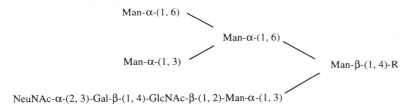

Figure 2.17

The growth of tumour cells treated in this way was reduced by 70% and metastasis was greatly reduced compared with normal tumour cells. The effect of alternative carbohydrate analogues on the growth and metastasis of tumours is highlighted in Table 2.3.

This demonstrates the potential of using carbohydrate analogues in cancer therapy, although again undesirable side-effects may be experienced with these agents due to their non-specific action.

Conclusions

The complex structures of many naturally occurring oligosaccharides makes their identification a challenging process, and consequently it is difficult to relate any particular biological process to a specific oligosaccharide. Although a multitude of biological processes can be attributed to the oligosaccharides, no single oligosaccharide displays all the biological properties associated with this group of molecules. Instead it would appear that oligosaccharides operate as a mixture of glycoforms. It is evident that there is still much progress to made in comprehending the roles of oligosaccharides and that exciting therapeutic applications may result once our understanding of their importance improves.

Table 2.3   Studies on MDAY-D2 lymphoma

Phenotype	Biological effect	Tumour size	Metastasis
Wild-type		4.55 ± 0.40 cm^3	> 500, 500, 500, 500
GlcNAc-transferase	Causes decrease in β-(1, 6)-GlcNAc branching	1.81 ± 0.39 cm^3	0, 3, 8, 8, 11
Swainsonine-treated	Competitive inhibitor of Golgi α-mannosidase II ($K_i = 40$ nm)	1.53 ± 0.50 cm^3	50–90% decrease.
UDP-Gal[†]	Loss of Gal and sialic acid from glycoproteins	0.10 ± 0.05 cm^3	0, 0, 0, 0, 0, 0
NeuNGlc[††]	Analogue of the naturally ocurring derivative of sialic acid, NeuNAc	$2.13 +/- 0.49$ cm^3	312, 400, 410, > 500, 500

[†] The UDP-Gal mutation is a defect in transport of this sugar into the Golgi
[††] *N*-Glycolylneuraminic acid

Progress in these areas will be greatly enhanced if access to large quantities of biologically important saccharides is made possible.

Further reading

1. Kornfeld, R., Kornfeld, S. (1985). Assembly of asparagine-linked oligosaccharides. *Annual Reviews in Biochemistry*, **54**, 631–664.

2. Feizi, T. (1993). Oligosaccharides that mediate mammalian cell–cell adhesion. *Current Opinion in Structural Biology*, **3**, 701–710.

3. Dwek, R.A. (1996). Glycobiology: Toward understanding the function of sugars. *Chemical Reviews*, **96**, 683–720.

4. Ogawa, T. (1994). Haworth memorial lecture. Experiments directed towards glycoconjugate synthesis. *Chemical Society Reviews*, **94**, 397–407.

5. Gabius, H-J. (1988). Tumor lectinology: At the intersection of carbohydrate chemistry, biochemistry, cell biology, and oncology. *Angewandte Chemie International Edition in English*, **27**, 1267–1276.

6. Favero, J. (1994). Lectins in AIDS research. *Glycobiology*, **4**, 387–396.

7. Look, G.C., Fotsch, C.H., Wong, C-H. (1993). Enzyme-catalysed organic synthesis: Practical routes to aza sugars and their analogs for use as glycoprocessing inhibitors. *Accounts of Chemical Research*, **26**, 182–190.

8. Lehmann, J. (translated by Alan H. Haines) (1998). 'Carbohydrates Structure and Biology', *Thieme Organic Chemistry Monograph Series*, Stuttgart, New York.

3   Synthesis of carbohydrate acceptors

General introduction to protecting-group strategies

Carbohydrate donors and acceptors are essential building blocks for all oligosaccharide syntheses. Indeed, in order to synthesise the wide range of oligosaccharides found in nature, a whole range of donors and acceptors are required which incorporate specific protecting-group patterns to allow entry only to the isomer of choice. For example, if the disaccharide shown below were required, an acceptor which displayed a free hydroxyl group only at C-4 would be required, together with a fully protected donor molecule.

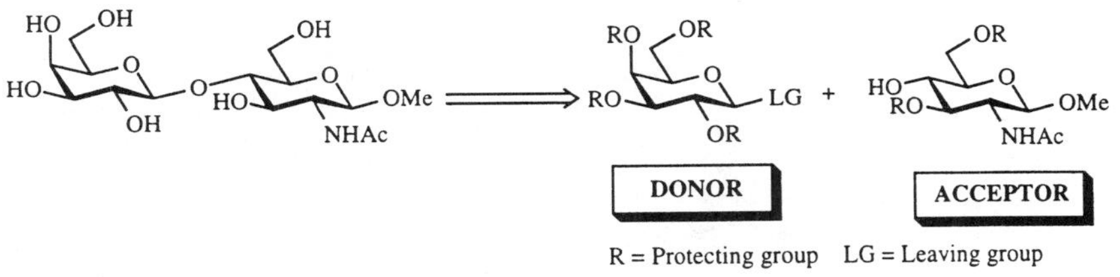

R = Protecting group   LG = Leaving group

Figure 3.1

In this chapter we will describe some chemical and enzymatic methods which allow synthesis of the partially protected acceptors which are commonly used in oligosaccharide syntheses. Protecting groups are employed in almost all chemical oligosaccharide syntheses and a whole range of useful, orthogonal protecting groups will be introduced in this chapter: orthogonal protecting groups are those which are introduced and removed under conditions which leave the remaining protecting groups intact.

For any synthetic strategy which involves protection/deprotection processes, it is essential to optimise the yields of all protection/deprotection steps so that the overall yield of the synthesis is not adversely affected. Furthermore, the selected protecting groups must be stable to any reaction conditions to which they may be exposed before the deprotection is to be effected. In most cases, protection and deprotection reactions which have become popular within carbohydrate-assembly strategies are indeed high yielding, and the range of orthogonal protecting groups available allows chemically compatible strategies to be developed. Since carbohydrate donors and acceptors often display similar orthogonal protecting groups they can normally be prepared using similar synthetic strategies.

Importance of protecting groups within carbohydrate-assembly strategies

To understand the importance of incorporating protecting groups within acceptor units, let us start by considering the synthesis of a range of disaccharides containing mannopyranoside units (Figure 3.2).

In order to synthesise the full range of disaccharide isomers, a whole range of acceptors is required. For example, an α-(1, 2) linked disaccharide would require an acceptor displaying a free hydroxyl group only at

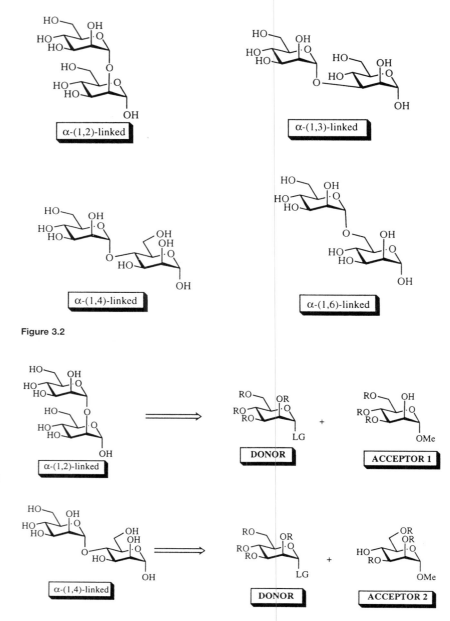

Figure 3.2

Protecting groups play important roles in chemical oligosaccharide synthesis.

Figure 3.3

C-2, whilst an α-(1, 4) linked disaccharide would require an acceptor displaying a free hydroxyl group only at C-4, as illustrated in Figure 3.3.

In order to synthesise all of the disaccharides depicted in Figure 3.2, we require entry to a range of acceptors which display a free hydroxyl group at any selected position. An indication of the protecting group patterns which must be incorporated within the acceptors is depicted in Figure 3.4. In the cases illustrated, a methoxy group is incorporated at the anomeric centre of the acceptor. This group is frequently incorporated within glycosyl acceptors since the methyl glycopyranosides required for the synthesis of the acceptors are commercially available in either anomeric form at economical rates. However, any anomeric group which is stable to the reaction conditions to be employed could be used.

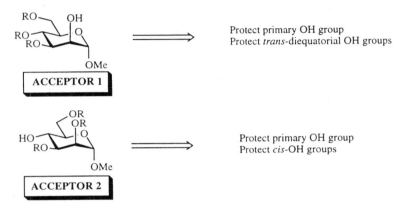

Figure 3.4

Some basic principles and methods which actually allow us to accomplish the protection patterns required are discussed below. In the laboratory we require access to more than just disaccharides containing mannopyranoside units: we must be able to elaborate the ideas presented to other pyranose and furanose units. Therefore some specific protecting groups which achieve selective protection are discussed below.

Specific protecting-group strategies

Protection of all the hydroxyl groups

There are a number of methods available in the literature for the total protection of all of the hydroxyl groups of a carbohydrate using commercially available reagents. For example, it is possible to protect all of the hydroxyl groups as acetate or benzoate esters using acetic anhydride or benzoyl chloride, respectively, in the presence of a base such as pyridine. If pyridine is used in excess it is also possible to employ it as the solvent for the re-action. In both cases, a catalytic quantity of 4-dimethylaminopyridine (DMAP) or imidazole can be utilised to increase the speed and efficiency of the re-action; the reaction mechanism for this process using DMAP with the monosaccharide methyl-α-D-mannopyranoside is highlighted in Figure 3.5.

Alternatively, all of the hydroxyl groups of the carbohydrate could be

Complete protection of the hydroxyl groups can be achieved by forming acetate or benzoate esters.

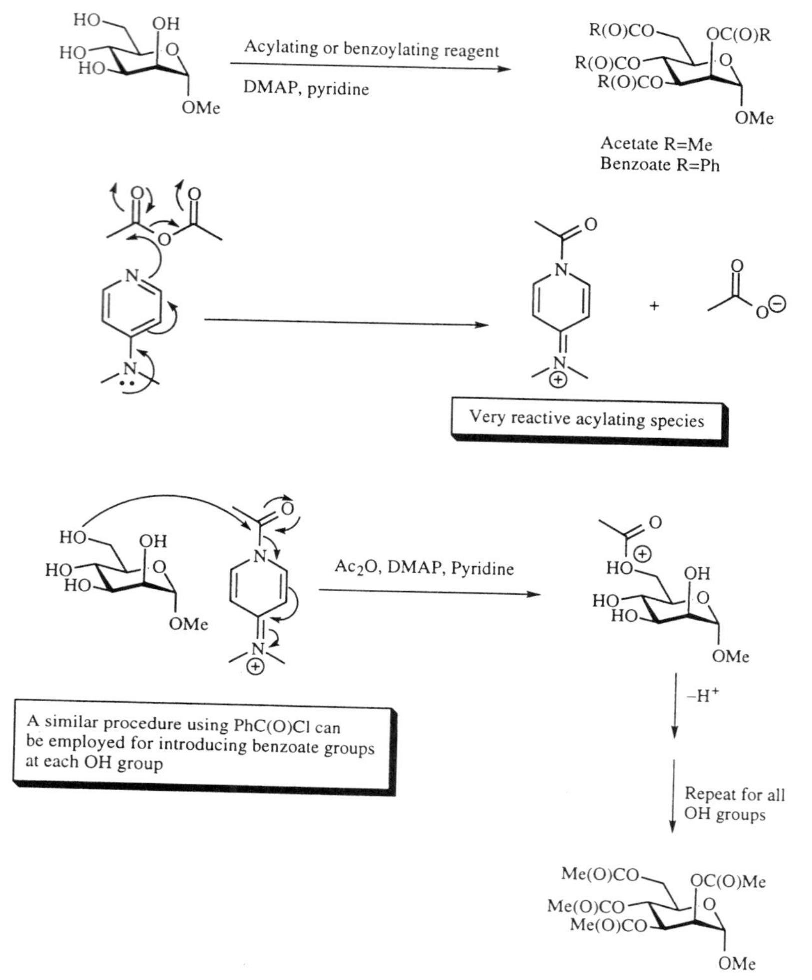

HO
OH
HO
HO
O
OMe
Acylating or benzoylating reagent
DMAP, pyridine
R(O)CO
OC(O)R
R(O)CO
R(O)CO
O
OMe
Acetate R=Me
Benzoate R=Ph
N
N
O
O
Very reactive acylating species
Ac2O, DMAP, Pyridine
–H+
Repeat for all OH groups
A similar procedure using PhC(O)Cl can be employed for introducing benzoate groups at each OH group
Me(O)CO
OC(O)Me
Me(O)CO
Me(O)CO
OMe

Figure 3.5

Complete protection of the hydroxyl groups can be achieved by forming benzyl ethers.

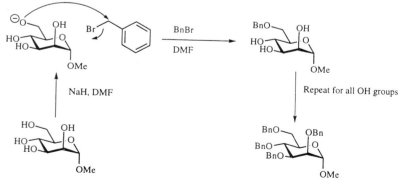

O
OH
HO
HO
OMe
Br
BnBr
DMF
BnO
OH
HO
HO
OMe
NaH, DMF
Repeat for all OH groups
HO
OH
HO
HO
OMe
BnO
OBn
BnO
BnO
OMe

Figure 3.6

protected as benzyl ethers by using the reagent combination of sodium hydride and benzyl bromide in a polar solvent such as dimethylformamide (DMF) as illustrated in Figure 3.6.

If these strongly basic reaction conditions are incompatible with the substrate undergoing protection then the transformation can instead be achieved under neutral conditions using benzyl triflate and 2,6–di-*tert*-butyl pyridine. The reaction conditions required for the introduction and removal of these protecting groups are noted in Table 3.1.

Table 3.1

Protecting group	Protection conditions	Deprotection conditions
Acetate (Ac)	Ac_2O, pyridine	NaOMe, MeOH
Benzoate (Bz)	BzCl, pyridine	NaOMe, MeoH
Benzyl (Bn)	BnBr, NaH, DMF	H_2, Pd/C

Deprotection of the acetate and benzoate esters involves initial attack by the methoxide anion on the ester carbonyl group. Collapse of the tetrahedral intermediate thus formed and expulsion of the sugar alkoxide then effects deprotection.

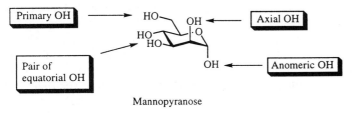

Figure 3.7

For the synthesis of specific acceptors we normally require access to partially protected molecules and hence a range of protecting groups which achieve partial protection are required.

Relative reactivity of hydroxyl groups: methods for selective protection of specific hydroxyl groups

If we examine a monosaccharide such as mannopyranose, we immediately notice the presence of one primary hydroxyl group at C-6 and four remaining secondary hydroxyl groups at C-1, C-2, C-3 and C-4.

The hydroxyl group at C-1 of free sugars exhibits different reactivity to those at C-2, C-3, C-4 and C-6 and will be examined separately in Chapter 4.

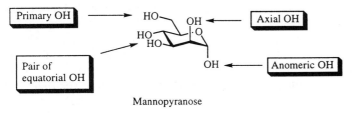

Mannopyranose

Figure 3.8

Selective protection of primary hydroxyl groups

Using basic chemical principals, we can predict that in the absence of external effects the primary hydroxyl group of an unprotected monosaccharide will be the most reactive hydroxyl group and it is therefore often possible to perform regioselective reactions with this C-6 hydroxyl group. Highlighted below are some typical protecting groups which can be chemically introduced onto this position in a regioselective manner. The yields and regioselectivities of these reactions are normally excellent, especially if sterically encumbered protecting groups are used.

The primary hydroxyl group is the most reactive and can be protected as a silyl or trityl ether.

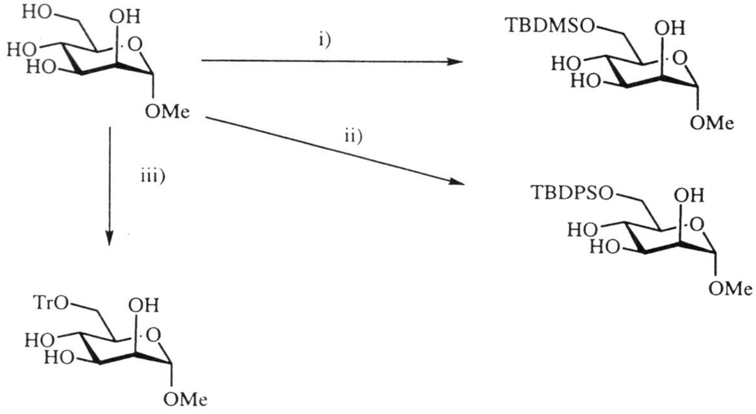

i) TBDMSCl, imidazole; ii) TBDPSCl, imidazole; iii) TrCl, imidazole

Figure 3.9

Hence selective protection of the C-6 hydroxyl group can typically be achieved by formation of a triphenylmethyl (trityl) or silyl ether at the C-6 position. These C-6 protected units provide useful intermediates for the synthesis of the acceptors depicted in Figure 3.4. A whole series of silyl ethers have been formed at the C-6 primary hydroxyl group, and these can be removed with varying ease depending on the size of the silyl ether. The actual group which is selected for use in a particular strategy depends upon the reaction conditions to be employed in subsequent steps, since the protecting group must remain intact until deprotection is required. In general, deprotection of silyl ethers can be achieved in excellent yield using a source of fluoride anions, for example tetrabutylammonium fluoride, or HF in pyridine, whilst acidic conditions are employed for removal of trityl ethers. Since the reactivity of fluoride anions decreases upon hydration, deprotection of silyl ethers using tetrabutylammonium fluoride can be improved if small quantities of activated molecular sieves are added to the reaction media.

Selective protection of the C-6 hydroxyl group of monosaccharides can also be achieved enzymatically (Figure 3.10). However, unprotected sugar substrates are highly polar and often require polar solvents for solubilisation. Unfortunately, many enzymes are inactive in such media so that problems are sometimes encountered. Porcine pancreatic lipase, PPL, is one enzyme which has proved suitable for effecting enzymatic transformation of unprotected monosaccharides. The protection reactions utilise

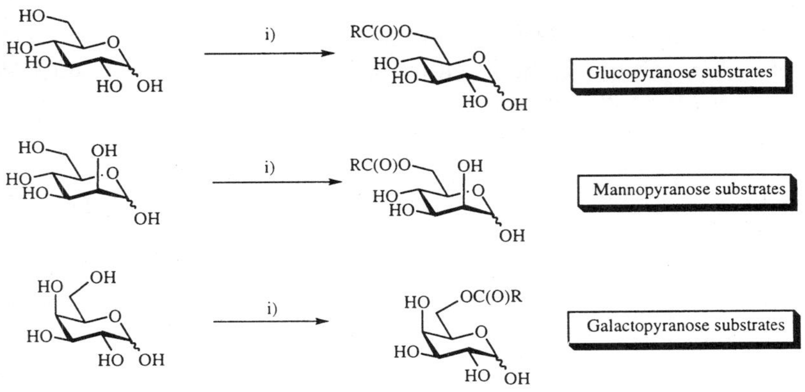

i) PPL, 2,2,2-trichloroethyl ester, pyridine

Figure 3.10

pyridine as the polar solvent and a variety of 2,2,2-trichloroethyl esters as the acylating agents. D-Glucopyranose, D-galactopyranose and D-mannopyranose are all substrates for this transformation, with yields and regioselectivities proving excellent.

The scope of these enzymatic protection reactions has been extended by using co-solvents and/or further reactive acylating agents, allowing entry to a range of pyranose and furanose building blocks in which the reactive primary hydroxyl group is protected.

Relative reactivities of remaining secondary hydroxyl groups

Some general rules concerning the relative reactivities of the remaining secondary C-2, C-3 and C-4 hydroxyl groups decorating carbohydrate rings can also be established. In general, an axial hydroxyl group is more reactive than an equatorial hydroxyl group and usually the C-2 hydroxyl group is more reactive than the C-3 hydroxyl group. This latter effect is due to the close proximity of the C-2 hydroxyl group to the anomeric centre. However, in the absence of extra reagents, it is generally difficult to effect regioselective reactions at any specific hydroxyl group other than the primary C-6 hydroxyl group, and mixtures of products often result under standard reaction conditions even if limiting quantities of protecting reagents are employed. Therefore methods for the selective protection of hydroxyl groups displaying particular stereochemical and regiochemical relationships have been developed, and these have advanced oligosaccharide synthesis strategies to a large extent.

Selective protection reactions

Protection of C-4 and C-6 hydroxyl groups as benzylidene acetals

The enhanced reactivity of the C-6 hydroxyl group compared with the remaining hydroxyl groups of pyranosides can be manipulated to allow simultaneous protection of the C-6 and C-4 hydroxyl groups. Since the C-4 hydroxyl group lies closest in space to the C-6 hydroxyl group,

The primary hydroxyl group can also be protected by the use of enzymes.

Axial hydroxyl groups are more reactive than equatorial hydroxyl groups.

simultaneous protection of both the C-4 and C-6 positions can be achieved if a reagent is introduced which is capable of producing a stable product by reaction at both centres. For example, use of benzaldehyde under acidic conditions (either protic or Lewis acid conditions) allows formation of a benzylidene acetal product.

The C-4 and C-6 hydroxyl groups can be protected by forming a benzylidene acetal.

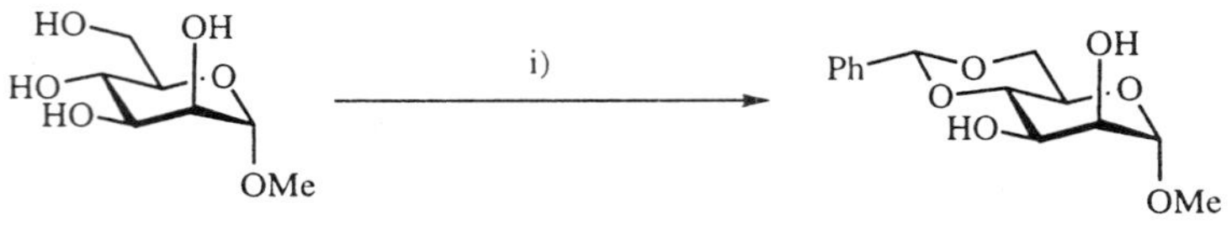

i) PhCHO, H+

Figure 3.11

This is particularly useful since the C-4 hydroxyl group is often the least-reactive secondary hydroxyl group of the pyranose ring and hence is difficult to selectively protect. The accepted mechanism for this reaction is depicted in Figure 3.12. A stable chair conformation is adopted by the newly formed six-membered acetal functionality and the aromatic ring is placed in the less-hindered equatorial position.

This reaction is not unique to mannopyranosides, but is equally useful for protection of alternative pyranosides such as gluco- and galactopyranosides. With galactose, however, yields of the analogous benzylidene product are slightly inferior due to formation of a different *stereoisomer* of the 4,6–benzylidene product, as well as formation of an alternate *regioisomer* resulting from reaction of the C-4 and C-3 hydroxyl groups (Figure 3.13). This is a result of the C-4 and C-3 hydroxyl groups of galactose occupying a *cis*-configuration, which allows them to form stable

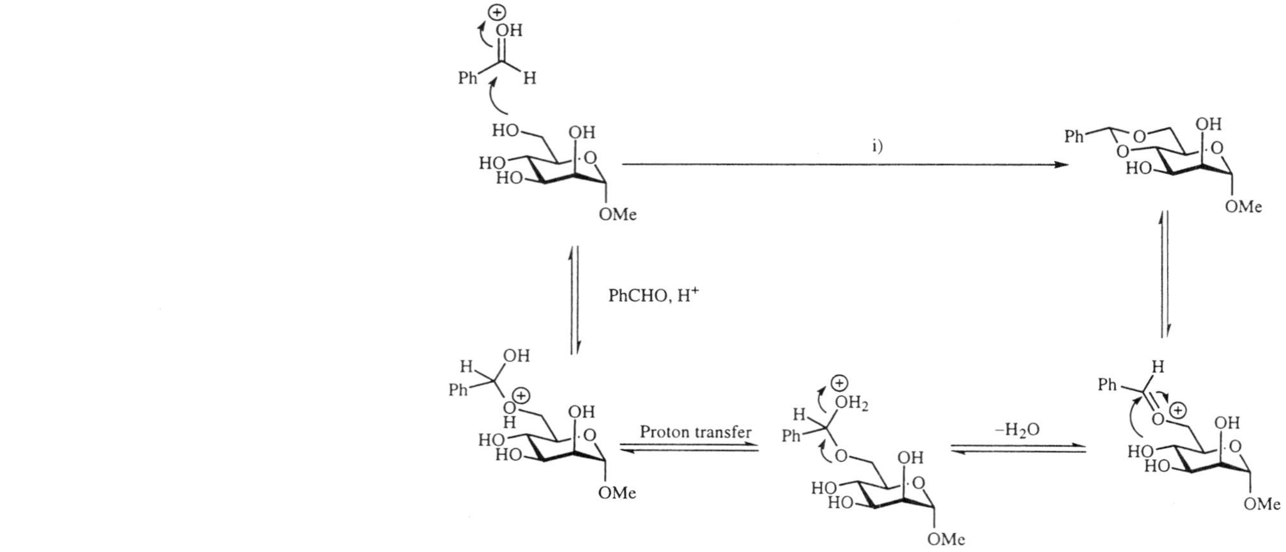

i) PhCHO, H+

Figure 3.12

A range of stereo- and regioisomers exist for benzylidene-protected glucopyranosides.

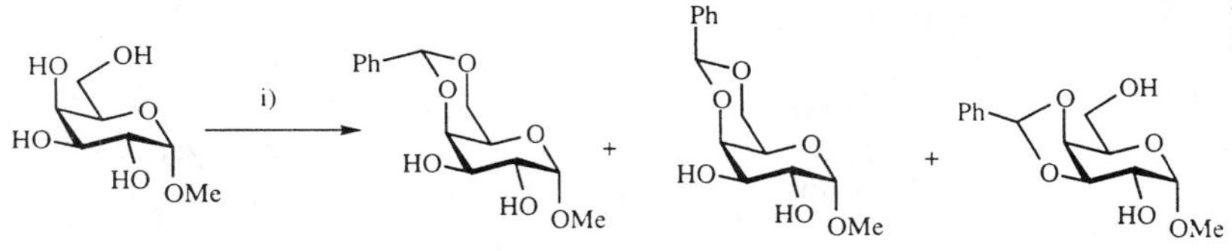

i) PhCHO, H+

Figure 3.13

acetal products. This reaction therefore competes to a small extent with the desired protection reaction of the C-4 and C-6 hydroxyl pair.

In all cases, benzaldehyde can be replaced in the protection reactions by benzaldehyde dimethyl acetal, which is essentially a more reactive form of benzaldehyde itself. An acid catalyst is still essential for formation of the key activated intermediate.

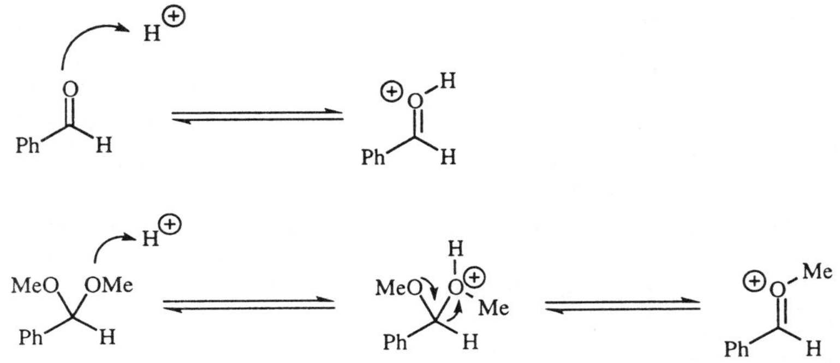

Figure 3.14

As highlighted in Figure 3.12, the formation of the benzylidene acetal is an equilibrium process which relies upon removal of water to drive the protection reaction to completion. Hence to reverse this process and effect deprotection, water must be added to the reaction. Indeed, deprotection of any acetal is typically achieved by treating the protected derivative with aqueous acid, as highlighted in Figure 3.15.

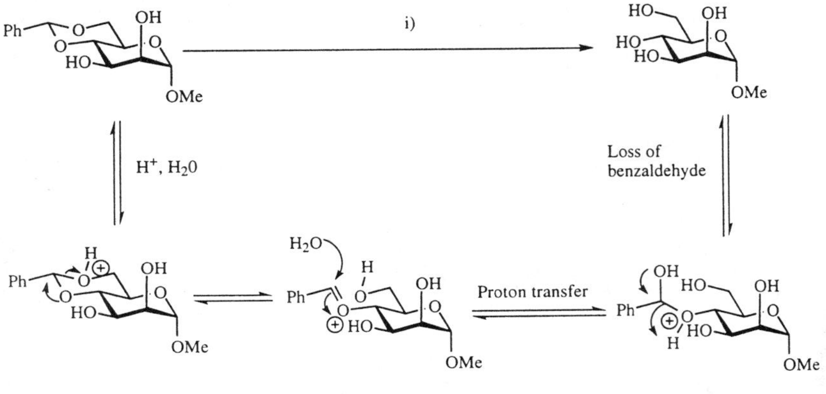

i) H_3O^+

Figure 3.15

Benzylidene acetals can be removed under acidic hydrolysis conditions, or by hydrogenolysis.

Excellent deprotection yields are achieved in this way. If, however, removal of the benzylidene acetal under acidic conditions is unsuitable due to the presence of other acid-sensitive moieties within the molecule, deprotection can instead be effected *via* catalytic hydrogenolysis employing hydrogen and Pd(OH)$_2$ as catalyst in ethanol.

Alternatively, a *p*-methoxybenzylidene acetal can serve as a protecting group for the hydroxyl groups at C-4 and C-6. Protection with this reagent is again achieved using a dimethyl acetal, but in this case deprotection can be achieved under oxidative conditions using 2,3-dichloro-5,6-dicyano-1,4-benzoquinone (DDQ), rather than acidic conditions. This is particularly useful for more sensitive molecules.

Benzylidene acetals also serve as useful precursors to other differentially protected products, and this will be dealt with in more detail at a later stage in this Chapter.

Protection of *cis*-hydroxyl groups

As highlighted earlier, acetal chemistry plays an important role in carbohydrate chemistry. A further example of a particularly useful acetal-protecting group is the acetonide-protecting group, which allows protection of any *cis*-pair of hydroxyl groups within a carbohydrate structure. The reagents required for this protection are acetone and an acid catalyst, and again protic and Lewis acid catalysts have each been successfully employed. The mechanism follows principles identical to those already illustrated for benzylidene acetal formation and is shown in Figure 3.16 for a mannopyranoside derivative with a *cis*-pair of hydroxyl groups at C-2 and C-3.

Interestingly, even if no *cis*-pair of hydroxyl groups is present in the sugar, interconversion of the pyranose and furanose forms may still allow protection to occur. For example, glucose displays *trans*-hydroxyl groups

Cis-hydroxyl groups can be protected by forming an acetonide.

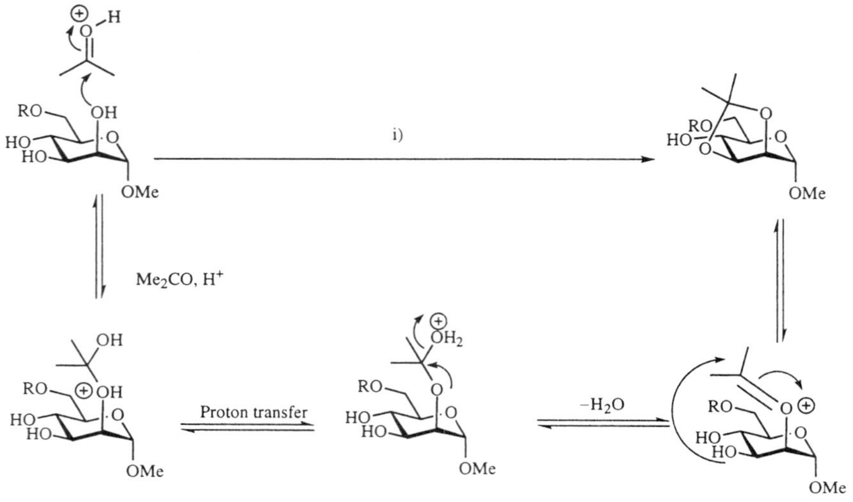

i) Acetone, H$^+$

Figure 3.16

only in its pyranose form, but once converted to its furanose form, hydroxyl groups in the *cis*-configuration are available which are able to react with acetone in an efficient manner to produce di-acetone glucose.

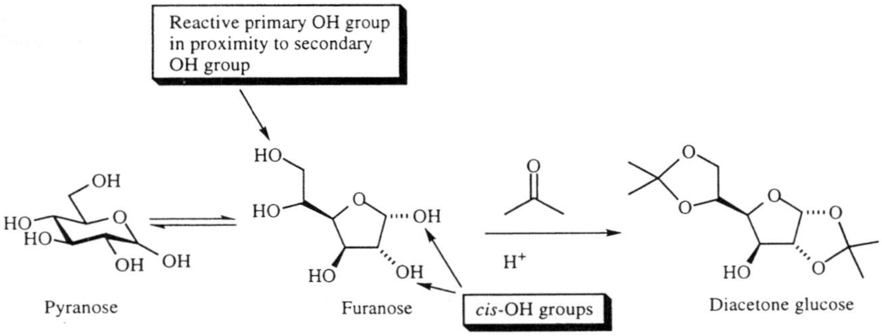

Figure 3.17

The selective protection of the *cis*-hydroxyl groups is therefore a well-established procedure which employs acetone and an acid catalyst. Deprotection of the acetal can be achieved under conditions identical to those described in Figure 3.15 for deprotection of benzylidene acetals, namely using aqueous acidic conditions.

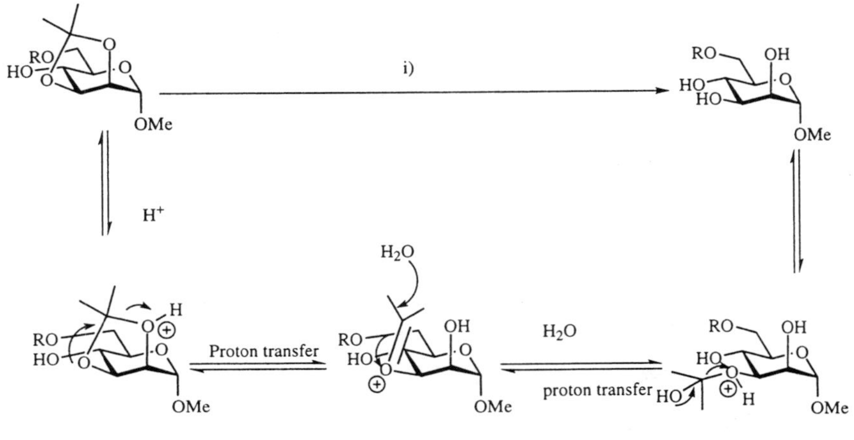

i) H_3O^+

Figure 3.18

Protection of *trans*-hydroxyl groups

Until recently, efficient selective protection of *trans*-hydroxyl groups within carbohydrates was impossible to achieve and instead multi-step strategies were typically employed to indirectly achieve this aim. Recent developments have removed this limitation and methods are now available which allow one-step protection of any *trans*-diequatorial pair of hydroxyl groups in monosaccharides. Two distinct classes of compounds have been developed to effect this protection, namely bis-dihydropyrans (bis-DHP) and a selection of 1,2-diketones (Figure 3.19).

Trans-hydroxyl groups can be protected using bis-DHP or 1,2-diketones.

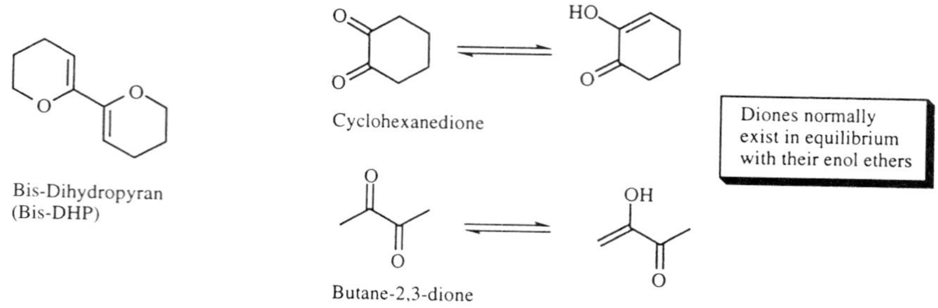

Bis-Dihydropyran
(Bis-DHP)

Cyclohexanedione

Butane-2,3-dione

Diones normally exist in equilibrium with their enol ethers

Figure 3.19

The diketones can instead be converted into their respective bis-dimethyl acetals prior to effecting protection since these bis-dimethyl acetals are also effective for protection of the *trans*-hydroxyl groups. The mechanisms for the formation of the reactive species from both the diketones and the bis-dimethyl acetals are shown in Figure 3.20.

The mechanism for protection of the *trans*-hydroxyl groups of a mannopyranoside using the reactive intermediate generated by the bis-dimethyl acetal is illustrated in Figure 3.21.

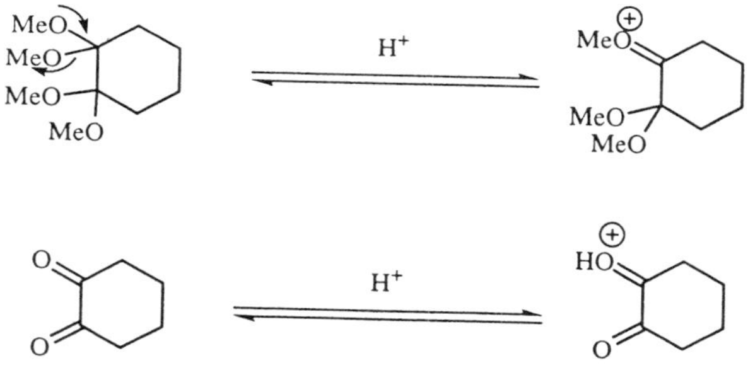

Figure 3.20

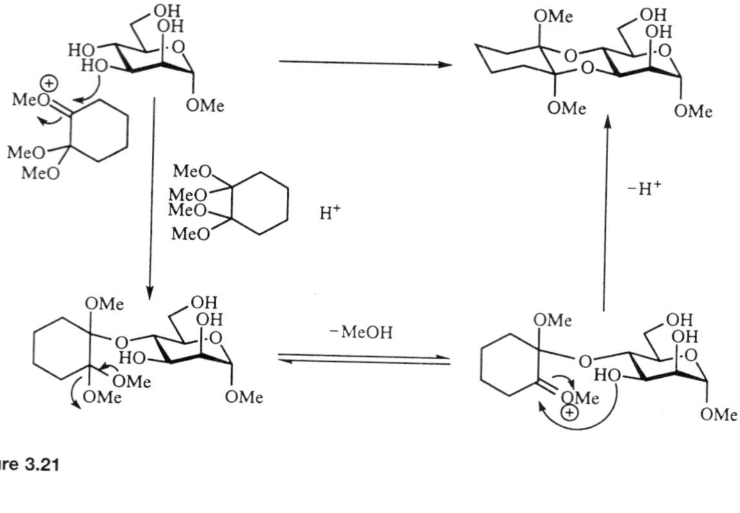

Figure 3.21

Bis-dihydropyran (bis-DHP) is also activated for protection under acidic conditions.

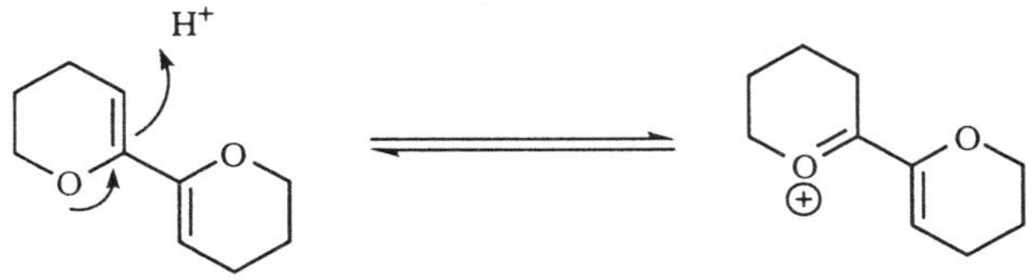

Figure 3.22

Initially, interception of this carbonium ion with either the C-3 or C-4 hydroxyl group of mannopyranoside occurs. A series of reactions then occurs to result in protection of the required *trans*-hydroxyl groups as illustrated in Figure 3.23.

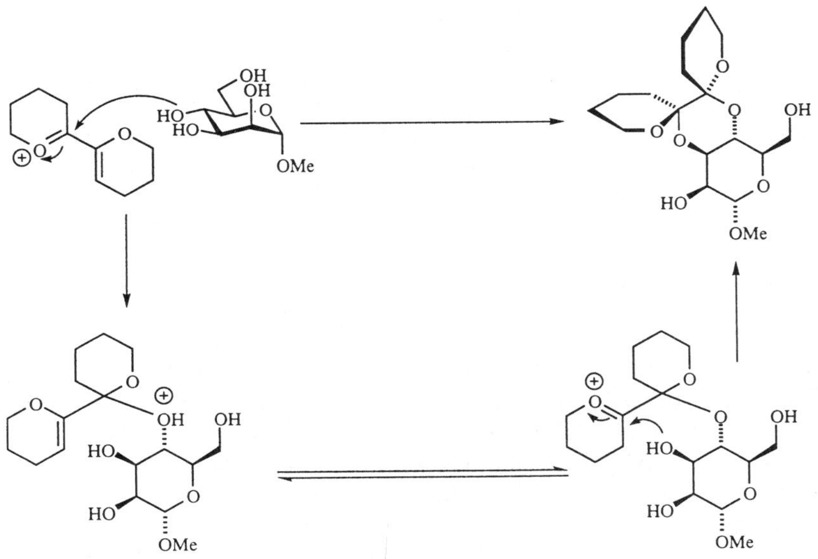

Figure 3.23

In both cases, selective protection of the *trans*-hydroxyl groups rather than the *cis*-hydroxyl groups results, and this is due to formation of extended cyclohexane chair conformations. The anomeric configuration of the newly formed acetal bonds is governed by the anomeric effect, and this phenomenon is discussed in detail in Chapter 5. Prior protection of the more-reactive C-6 hydroxyl group is not a prerequisite for the efficiency of this reaction, but it is sometimes advantageous due to the poor solubility of unprotected carbohydrates in most organic solvents. Deprotection of the protected compounds can again be achieved under acidic conditions following mechanisms analogous to those in Figure 3.18. If acidic conditions are incompatible with functionality within the sugar moiety, then a *bis*-dihydropyran incorporating alternative functionality may be employed. Some modified *bis*-dihydropyrans which have been shown to be removable under alternative chemical conditions are highlighted in Figure 3.24 and Table 3.2.

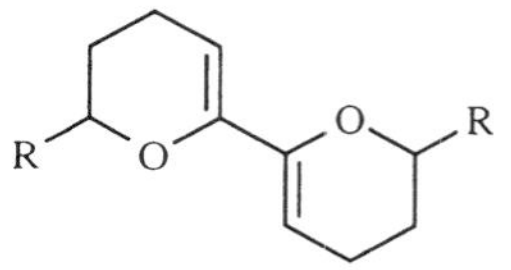

Figure 3.24

Table 3.2

Nature of substituent, R	Deprotection conditions
Ph	Excess iron (III) chloride, room temperature overnight
Me	95% Trifluoroacetic acid (TFA), room temperature, 4 hours
Br	Lithium 4,4′-di-*t*-butylbiphenyl, LDBB (via reductive β-elimination)
I	As for Br. Also via treatment with Rieke Zn
PhS	*m*CPBA then lithium hexamethyldisilazide (LHMDS), THF, 0°C, 30 minutes (via oxidation to sulfone then β-elimination)
Allyl	O_3,–78°C then PPh_3 followed by diazabicyclo[5.4.0] under-7-ene (DBU), 80°C, 21 hours (via ozonolysis to aldehyde followed by β-elimination)

On first sight it may be thought that glucopyranosides which display two pairs of relatively non-sterically differentiated enantiomeric *trans*-hydroxyl groups may be unsuitable for protection using the dispiroketal or dione protecting groups highlighted above. Indeed, if glucopyranoside is treated with *bis*-DHP or the diones under typical acid reaction conditions, two products derived from protection of the C-2 and C-3 or the C-3 and C-4 hydroxyl pairs result in essentially equal amounts. Hence no distinction is made between the *trans*-hydroxyl groups (Figure 3.25).

However, it has proved possible to overcome this limitation by developing chiral versions of bis-DHP which match the enantiotopicity of either the C-2 and C-3 *trans*-hydroxyl pair or the C-3 and C-4 *trans*-hydroxyl pair. For example, the (*R, R*) enantiomer of the bis-phenyl

Glucopyranosides have two pairs of *trans*-hydroxyl groups.

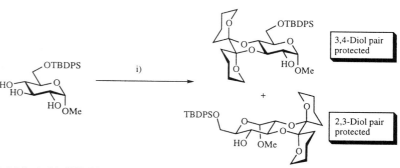

3,4-Diol pair protected

2,3-Diol pair protected

i) 2.1 Equiv. bis-DHP, CSA, $CHCl_3$, heat 1.5 hours then ethylene glycol

Figure 3.25

derivative of bis-DHP matches the enantiotopicity at C-2 and C-3 of D-glucopyranoside substrates, whilst the opposite (*S*, *S*) enantiomer matches the enantiotopicity at C-3 and C-4 of D-glucopyranoside substrates.

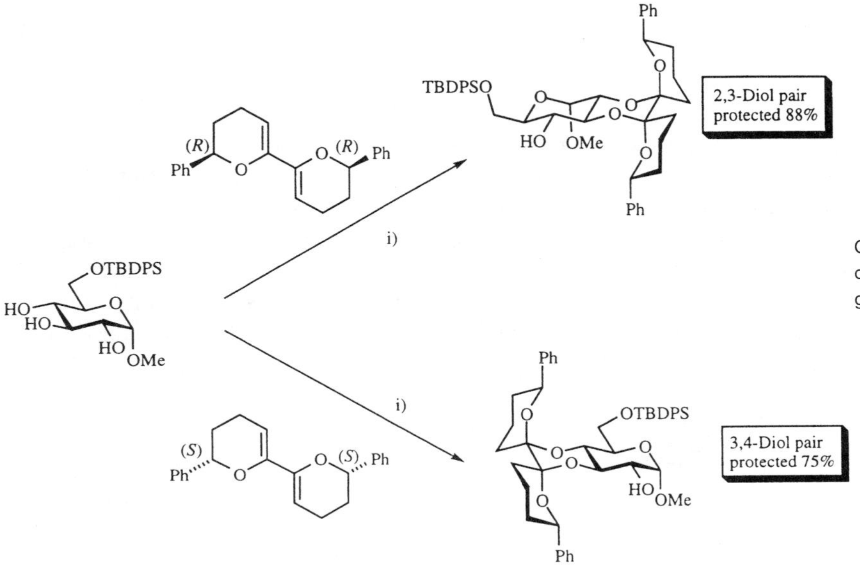

Chiral variants of bis-DHP allow protection of either pair of *trans*-hydroxyl groups in glucopyranosides.

i) 2.1 Equiv. bis-DHP, CSA, CHCl₃, heat 1.5 hours then ethylene glycol

Figure 3.26

We have almost discussed a sufficient number of reactions to allow synthesis of the acceptors depicted in Figure 3.3. Hence protection of the C-6 hydroxyl group followed by protection of the *trans*-hydroxyl groups affords the acceptor necessary for synthesis of the α-(1,2)-linked disaccharide, whilst protection of the C-6 hydroxyl group followed by protection of the *cis*-hydroxyl groups affords the acceptor necessary for synthesis of the α-(1,4)-linked disaccharide. However, in order to synthesise the acceptor necessary for entry to the α-(1,3)-linked disaccharide, we need a method of protecting the hydroxyl groups at C-2, C-4 and C-6, leaving the C-3 hydroxyl group free. One particularly useful way of achieving this is to perform further reactions on monosaccharide benzylidene acetals in which the C-2 and C-3 hydroxyl groups remain unprotected. This has been achieved in a number of ways.

Further elaboration of benzylidene acetals

Metal chelate formation

Reports in the literature have described the use of tin acetals for further selective protection of monosaccharide benzylidene acetals. Chelation of benzylidene-protected *glucopyranoside* diols with organotin complexes allows selective activation of the hydroxyl group at C-2 to subsequent protections.

Even better reaction yields can be attained when the more reactive stannylene derivative dibutyltin dimethoxide (Bu₂Sn(OMe)₂) is employed for chelate formation under Dean–Stark conditions. The exact mode of action of the stannylene-activating mechanism is not clear, but a hypothesis

Organotin complexes allow selective protection of benzylidene-protected glucopyranosides.

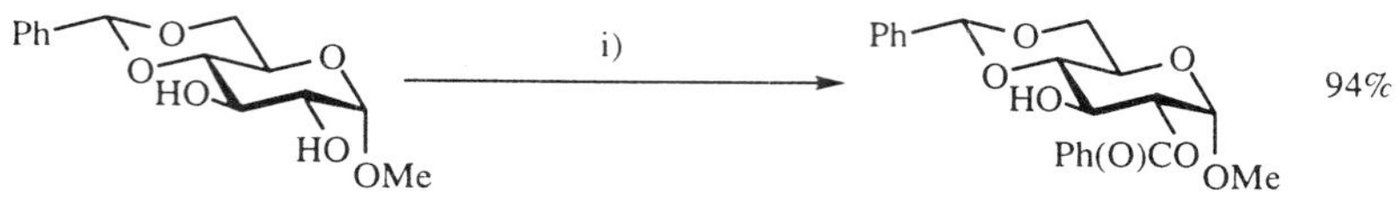

i) Bu$_2$Sn(OMe)$_2$, BzCl

Figure 3.27

has been proposed which suggests that one of the two oxygen atoms involved in the stannylene ring formation becomes more nucleophilic due to electron channelling from the tin atom.

Alternatively, copper chelates can be used to afford the opposite regioisomer to that obtained using the tin acetal methodology, that is, entry to acceptors with a free hydroxyl group at C-2. In these cases, sodium hydride is employed to form the benzylidene dialkoxide, which is treated with anhydrous copper (II) chloride to form a copper chelate.

Organocopper complexes allow selective protection of benzylidene-protected glucopyranosides.

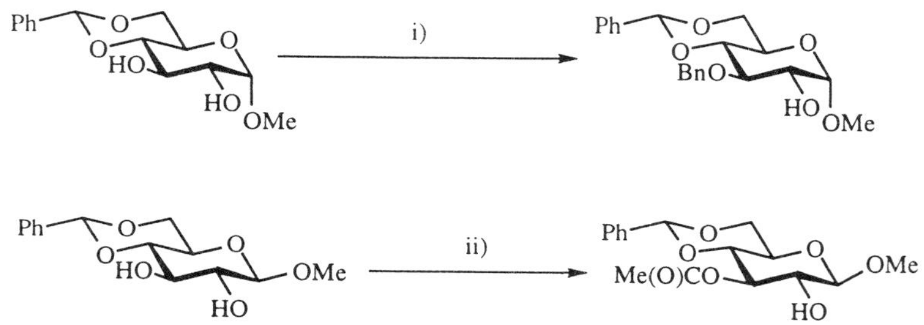

i) 2 equiv NaH, CuCl$_2$, BnBr; ii) 2 equiv NaH, CuCl$_2$, AcCl

Figure 3.28

In these cases it is postulated that formation of the copper chelate de-activates the dianion from which it is derived, making both the hydroxyl groups at C-2 and C-3 less reactive. Since the hydroxyl group at C-2 is intrinsically more reactive than that at C-3, due to its proximity to the anomeric centre, it binds most strongly to the copper ion. The hydroxyl group at C-3 is therefore more free for subsequent reactions with acylating or alkylating agents and it is this position which becomes substituted in the subsequent reaction. Interestingly, both the α- and β-anomers of the benzylidene-protected monosaccharide acceptors are reported to allow regioselective entry to C-2 acceptors as highlighted above for the glucopyranoside substrates.

Enzyme-mediated reactions

Regioselective protection of benzylidene-protected monosaccharides can also be achieved using acylating agents in conjunction with a range of commercially available enzymes, and the reactions are usually performed in THF. The range of enzymes investigated includes Lipase P (from *Pseudomonas cepacia* adsorbed on Celite® to improve reproducibility), *Pseudomonas fluorescens* lipase (PFL) and *Candida cylindracea*. Although a range of monosaccharides have been investigated, the most useful results have been obtained with gluco- and mannopyranosides (Figure 3.29).

Enzymes allow selective protection of benzylidene-protected glucopyranosides.

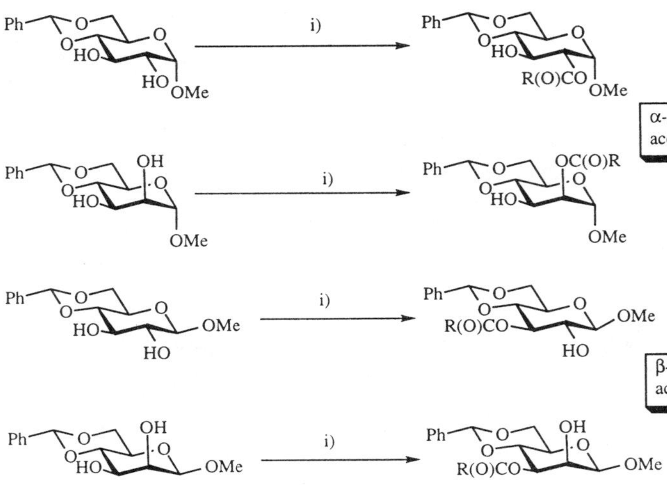

α-Anomer affords C-3 acceptor

β-Anomer affords C-2 acceptor

i) Enzyme, $H_2C=CHOC(O)R$

Figure 3.29

It has been shown that the anomeric configuration of the monosaccharide influences the regioselectivity of the reaction, with the α-anomer affording the C-3 acceptor and the β-anomer affording the C-2 acceptor. Regioselectivities and yields are typically excellent, allowing efficient preparation of gluco- and mannopyranoside acceptors. With other monosaccharide substrates, such as methyl 4,6-*O*-benzylidene-α-D-galactopyranoside, yields of products are lower and the reactions are slower. However, even when yields of products are low, excellent regioselectivities which mirror those illustrated for the gluco- and mannopyranoside substrates are obtained. These enzyme-mediated procedures therefore allow entry to the specific acceptors required for formation of C-2- or C-3-linked oligosaccharides.

Chemical manipulation

A further important use of benzylidene-protected monosaccharides which must be mentioned is their manipulation to afford either C-4 or C-6 acceptors. For example, it has been shown that if a monosaccharide benzylidene acetal is treated with aluminium chloride and lithium aluminium hydride, the C-6 acceptor with a benzyl ether at C-4 is obtained in excellent yield. However, entry to the C-4 acceptor with a benzyl ether at C-6 can alternatively be achieved by treatment of the benzylidene acetal with sodium cyanoborohydride in THF using dry HCl as catalyst (Figure 3.30).

At this stage it is useful to provide a summary of the methods we have discussed which allow selective protection to be achieved. We present this in Table 3.3. A summary of conditions effective for deprotection of compounds displaying only one type of protecting group is also provided.

We are now in a position to return to Figure 3.2 and devise plausible syntheses of the acceptors required for entry to the disaccharide targets. Combinations of the protecting groups described above are employed as shown in Figure 3.31. For example, the acceptor required for entry to the α-(1,2)-linked disaccharide could be prepared by first protecting the

Treatment of benzylidene acetals with AlCl$_3$ and LiAlH$_4$ affords C-6 acceptors whilst treatment with NaBH$_3$CN and HCl affords C-4 acceptors.

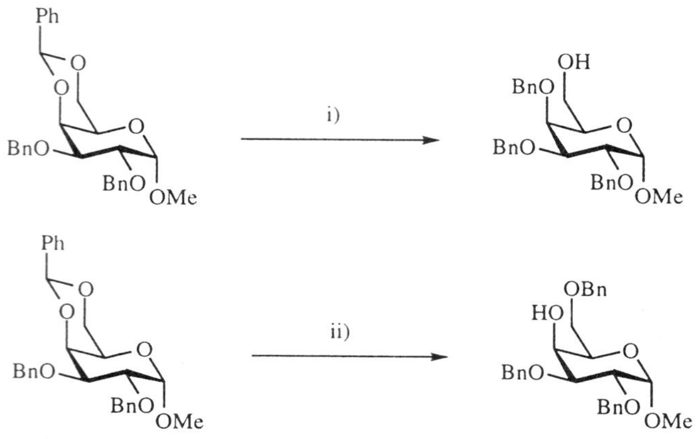

i) AlCl$_3$, LiAlH$_4$; ii) NaBH$_3$CN, HCl

Figure 3.30

most reactive C-6 hydroxyl group as its silyl ether and subsequently protecting the *trans*-hydroxyl groups with butane dione under acid catalysis. On the other hand, the acceptor required for entry to the α-(1,4)-linked disaccharide would require protection of the C-6 hydroxyl group as its silyl ether followed by protection of the *cis*-hydroxyl groups with acetone and acid. The acceptor required for entry to the α-(1,3)-linked disaccharide would require protection of the hydroxyl groups at C-4 and C-6 as a benzylidene acetal, followed by protection of the more-reactive axial hydroxyl group at C-2 as an acetate or benzoate ester, or as a benzyl ether. Finally, the acceptor required for entry to the α-(1,6)-linked disaccharide could be prepared in a number of ways. For example, a benzylidene acetal could be temporarily employed to simultaneously protect the hydroxyl groups at C-4 and C-6. Further protection of the remaining diol pair would then be required prior to selective opening of the benzylidene acetal to afford the C-6 acceptor. Alternatively, the C-6 hydroxyl group could be protected as a silyl ether. Complete protection of the remaining hydroxyl groups as their acetate esters or benzyl ethers would then afford the fully protected compound. Upon treatment with tetrabutylammonium fluoride, selective deprotection of the silyl ether would occur to afford the required acceptor.

Synthesis of branched-chain oligosaccharides

The above strategies have described synthesis of acceptors displaying only one free hydroxyl group. Although this class of acceptors is widely utilised in oligosaccharide syntheses, they form only one class of acceptors currently employed. An alternative class of acceptors which are often employed for the synthesis of branched oligosaccharides bear more than one free hydroxyl group. For example, if the branched oligosaccharide in Figure 3.32 were required, an acceptor with two free hydroxyl groups would be necessary.

The range of protecting groups introduced in this Chapter are still of use for the synthesis of this molecule. For example, the acceptor illus-

Table 3.3

Hydroxyl groups protected	Reagents effecting protection	Reagents effecting deprotection
Complete protection of all OH groups	Acetic anhydride, pyridine	Sodium methoxide, methanol
Complete protection of all OH groups	Benzoyl chloride, pyridine	Sodium methoxide, methanol
Complete protection of all OH groups	Benzyl chloride, sodium hydride, DMF	Hydrogen, palladium catalyst
Protection of C-6 OH group	TBDMSCl, DMF, imidazole	TBAF or HF/pyridine
Protection of C-6 OH group	Trityl chloride, DMF	Acid
Selective protection of C-4 and C-6 OH groups	Benzaldehyde, acid or benzaldehyde dimethyl acetal, acid	Acid, water or hydrogen/ palladium catalyst
Selective protection of *cis*-OH groups	Acetone, acid	Acid, water
Selective protection of *trans*-OH groups	Butane dione, acid	Acid, water
Selective protection of *trans*-OH groups	Bis-dihydropyran, acid	Acid, water
Selective protection of C-3, C-4 and C-6 OH groups	Benzaldehyde, acid to form benzylidene acetal then $Bu_2Sn(OMe)_2$, alkylating agent	
Selective protection of C-3, C-4 and C-6 OH groups	Benaldehyde, acid to form benzylidene acetal then acylating agent (e.g. vinyl acetate), PFL enzyme	
Selective protection of C-2, C-4 and C-6 OH groups	Benzaldehyde, acid to form benzylidene acetal then 2 equiv. NaH, $CuCl_2$, acylating or alkylating agent	
Selective protection of C-6 OH group	Benzaldehyde, acid to form benzylidene at C-4 and C-6 then sodium cyanoborohydride, HCl, THF	
Selective protection of C-4 OH group	Benzaldehyde, acid to form benzylidene at C-4 and C-6 then $LiAlH_4$, $AlCl_3$	

trated in Figure 3.32 could be accessed from the benzylidene acetal of methyl-α-D-mannopyranoside using an enzyme-mediated acylation reaction to initially protect the C-2 hydroxyl group. Ring opening of the benzylidene acetal with lithium aluminium chloride and aluminium chloride would then afford the required acceptor (Figure 3.3).

Two equivalents of a donor molecule could then react with this acceptor to afford the required trisaccharide.

In this chapter we have concentrated on the synthesis of acceptor

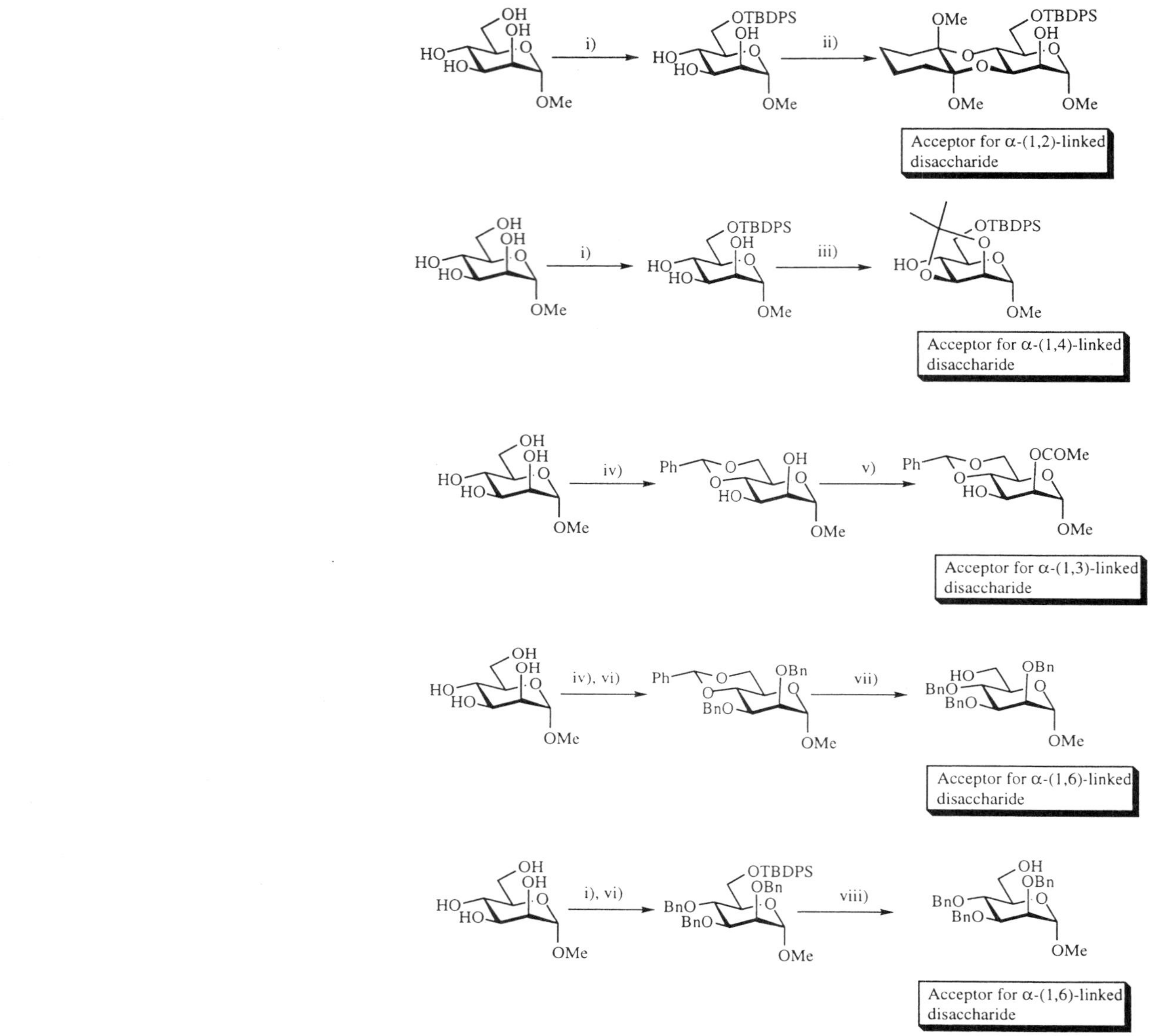

i) TBDPSCl, imidazole; ii) Cyclohexane dione, H$^+$, (MeO)$_3$CH, MeOH; iii) Acetone, H$^+$; iv) PhCho, H$^+$; v) vinyl acetate, PFL; vi) BnBr, NaH; vii) LiAlH$_4$, AlCl$_3$; viii) TBAF

Figure 3.31

molecules as useful precursors to oligosaccharides. We have made only passing comments on the synthesis of the donor components which are also essential for the syntheses. Therefore in the next chapter we will examine the synthesis of carbohydrate donors.

Further reading

1. Garegg, P.J. (1984). Some aspects of regio-, stereo-, and chemoselective reactions in carbohydrate chemistry. *Pure and Applied Chemistry*, **56**, 845–858.

Branched-chain oligosaccharides require acceptors with more than one free hydroxyl group.

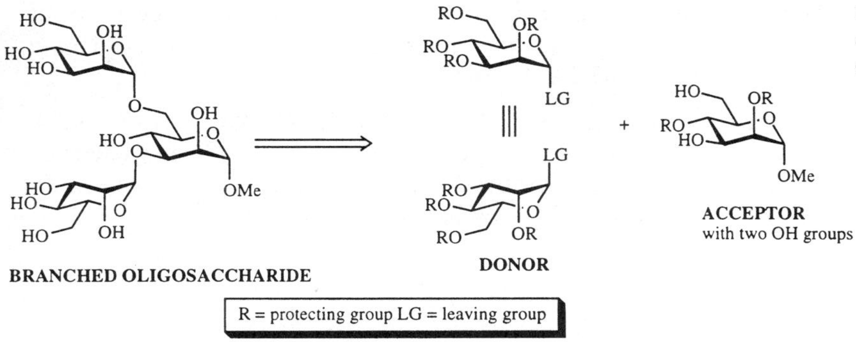

R = protecting group LG = leaving group

Figure 3.32

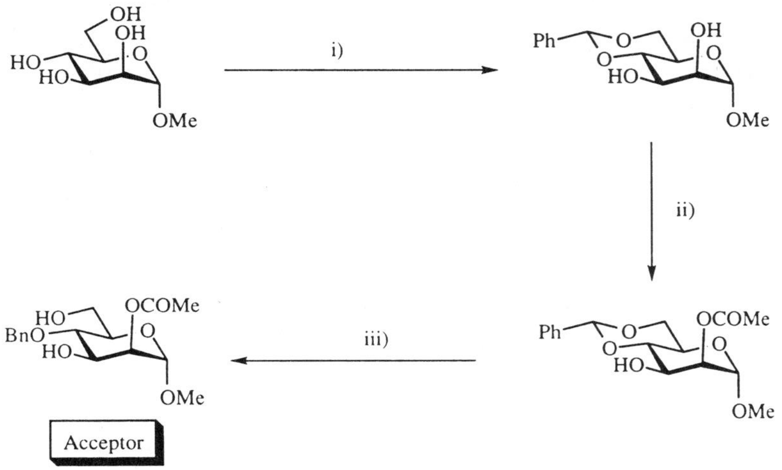

i) PhCHO, H⁺; ii) vinyl acetate, PFL; iii) LiAlH₄, AlCl₃

Figure 3.33

2. Ley, S.V., Priepke, H.W.M., Warriner, S.L. (1995). Cyclohexane-1,2-diacetals (CDA): A new protecting group for vicinal diols in carbohydrates. *Angewandte Chemie International Edition in English*, **33**, 2290–2292.

3. David, S., Hanessian, S. (1985). Regioselective manipulation of hydroxyl groups via organotin derivatives. *Tetrahedron*, **41**, 643–663.

4. Grindley, T.B. (1998). Applications of tin containing intermediates to carbohydrate chemistry. *Advances in Carbohydrate Chemistry and Biochemistry*, **53**, 17–142.

5. Roberts, S.M. (1998). Preparative biotransformations : the employment of enzymes and whole cells in synthetic organic chemistry. *Journal of the Chemical Society, Perkin Transactions I*, 157–169.

6. Bashir, N.B., Phythian, S.J., Reason, A.J., Roberts, S.M. (1995). Enzymatic esterification and de-esterification of carbohydrates—synthesis of a naturally occurring rhamnopyranoside of *p*-hydroxybenzaldehyde

and a systematic investigation of Lipase-catalysed acylation of selected arylpyranosides. *Journal of the Chemical Society, Perkin Transactions I*, 2203–2222.

7. Gridley, J.J., Hacking, A.J., Osborn, H.M.I., Spackman, D.G. (1998). Regioselective Lipase-catalysed acylation of 4,6–*O*-benzylidene-α- and β-D-pyranoside derivatives displaying a range of anomeric substituents. *Tetrahedron,* **54**, 14925–14946.

4 Synthesis of carbohydrate donors

Together with carbohydrate acceptors, carbohydrate donors are essential molecules for the synthesis of oligosaccharides. Carbohydrate donors incorporate a group at C-1 which can be displaced under specific reaction conditions by carbohydrate acceptors. This process often involves transient formation of a carbonium ion which is subsequently trapped by the free hydroxyl group of the carbohydrate acceptor to allow formation of the new glycosidic bond.

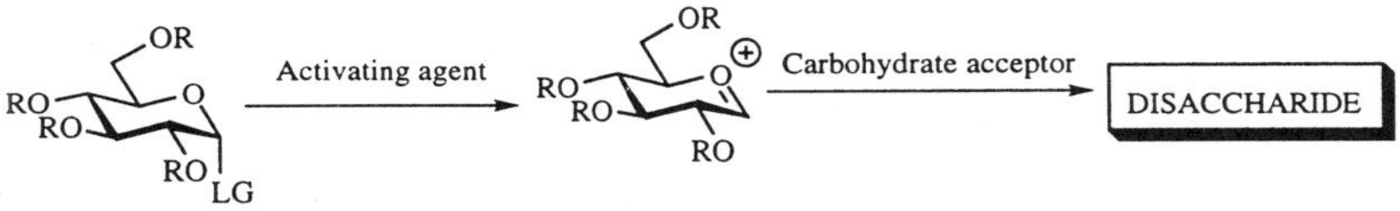

Carbohydrate donors incorporate a group at C-1, which can be displaced under specific reaction conditions by carbohydrate acceptors.

Figure 4.1

All of the remaining hydroxyl groups surrounding the donor are typically protected to inhibit any undesired competition with the acceptor molecule: this would lead to the formation of mixtures of oligosaccharides, which is of limited use if entry to a specific oligosaccharide of choice is required. Whether all the hydroxyl groups of the donor are protected by the same group, or by orthogonal groups, largely depends on the subsequent manipulations which may be required to complete the synthesis of the chosen oligosaccharide.

Common groups which are incorporated at the anomeric centre of carbohydrate donors include acetate, halides, thiols, selenides, trichloroacetimidate, n-pentenyl glycoside, vinyl glycoside, sulfoxides and glycals (Figure 4.2).

All of these anomeric groups have proved suitable for subsequent activation and displacement with an acceptor unit to form glycosidic linkages. Although glycals do not possess a leaving group at their anomeric position, they can still be activated to react with carbohydrate acceptors by prior reaction with electrophilic species and are hence still classified as carbohydrate donors. The donor which is actually used in a synthesis largely depends on the compatability of its anomeric group with the reaction conditions to which it will be exposed during the synthesis, as well as the compatability of the remaining protecting groups on the donor and acceptor to the activating agent required for formation of the carbonium ion.

Glycosyl acetates and thioglycosides

Glycosyl acetates and thioglycosides are extremely popular carbohydrate donors, and this is reflected, for example, by the fact that many mono- and

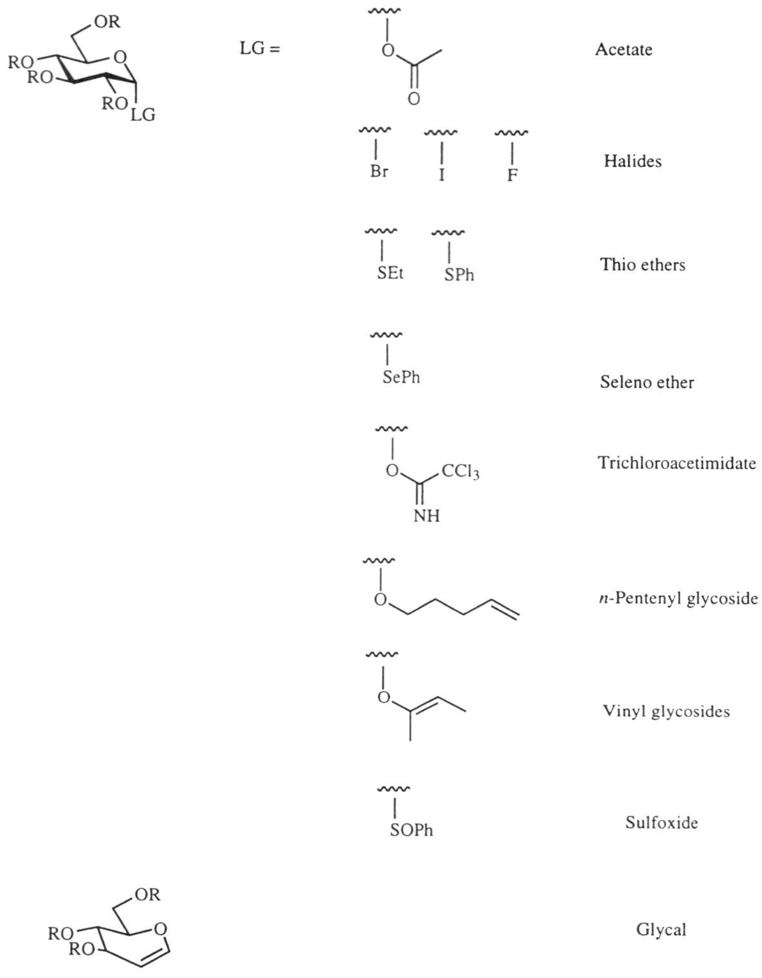

Figure 4.2

di-saccharide acetate donors are commercially available. For simplicity, the remaining hydroxyl groups of the donor are often themselves protected as acetate esters. The acetate group at C-1 can be activated for reaction with acceptors by treatment with a Lewis acid such as $BF_3.OEt_2$. This allows ready formation of the required carbonium intermediate (Figure 4.3).

Low temperatures of 0°C are often employed for such reactions, and solvents such as dichloromethane have proved particularly useful. The carbonium ion formed can be trapped either with carbohydrate acceptors, allowing formation of oligosaccharides, or by a number of alternative nucleophiles, which allow entry to other donors in good yields. For example, activated glycosyl acetates have been treated with thiols to allow synthesis of useful thioethyl glycosyl donors (Figure 4.4).

Thioglycosides are themselves useful donors since they are prone to activation with thiophilic reagents such as *N*-iodosuccinimide and triflic acid (NIS/TfOH), *N*-bromosuccinimide (NBS) and Hg^{II} salts. One example of activation using NIS/TfOH as the source of iodonium cations is highlighted below (Figure 4.5).

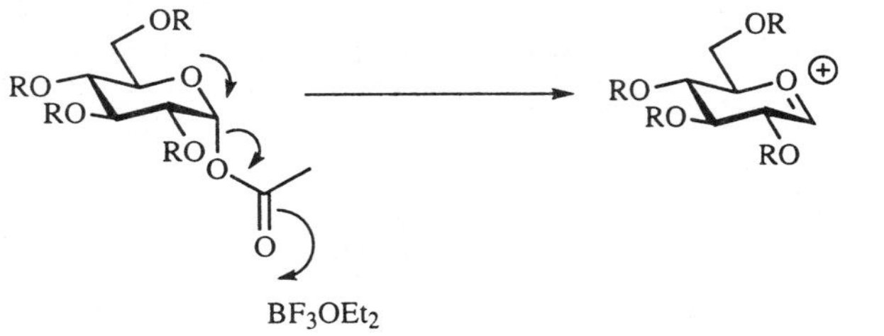

Glycosyl acetates can be activated with a Lewis acid such as BF$_3$.OEt$_2$.

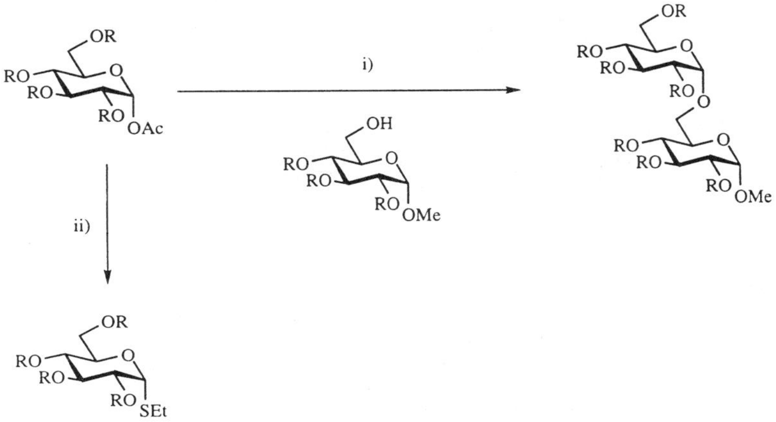

Figure 4.3

Thioglycosides can be prepared by treating glycosyl acetates with thiols in the presence of BF$_3$.OEt$_2$.

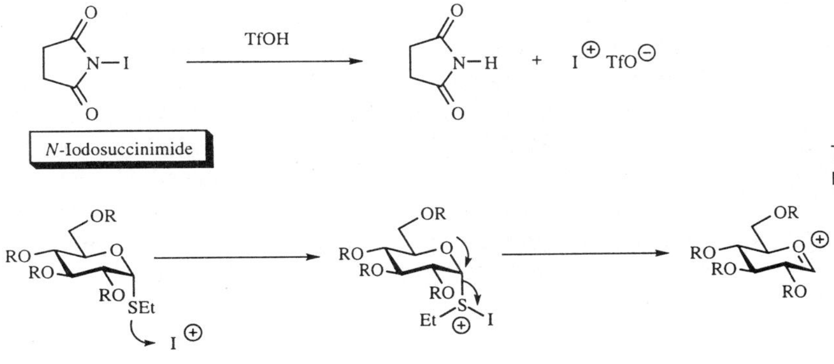

i) BF$_3$.OEt$_2$; ii) BF$_3$.OEt$_2$, EtSH

Figure 4.4

N-Iodosuccinimide

Thioglycosides can be activated with NIS/TfOH, NBS or Hg^{II} salts.

Figure 4.5

The carbonium ion thus formed is normally trapped with a carbohydrate acceptor, allowing entry to larger oligosaccharides. Interconversion of the thiopyranoside to alternative glycosyl donors such as the analogous sulfoxide, fluoride or bromide is also often encountered in carbohydrate chemistry. For example, oxidation of the thioglycosides with *m*-chloroperbenzoic acid (*m*-CPBA) affords the sulfoxide, whilst treatment with *N*-bromosuccinimide (NBS) and diethylaminosulfur trifluoride

(DAST) generates the glycosyl fluoride. Treatment with bromine can be used to convert the thioglycoside to the glycosyl bromide.

Thioglycosides can be converted to glycosyl sulfoxides, glycosyl fluorides or glycosyl bromides by treatment with *m*-CPBA or Oxone® , NBS/DAST and bromine, respectively.

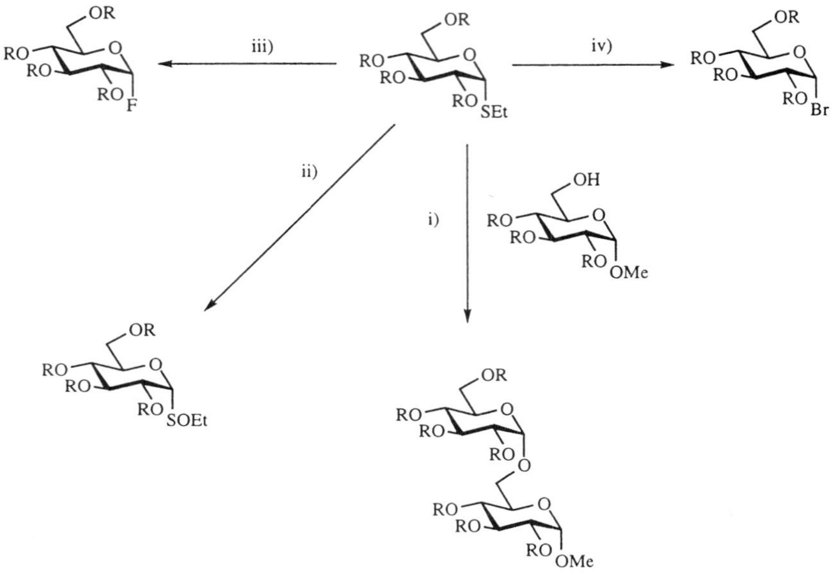

i) NIS, TfOH; ii) *m*-CPBA, –20°C; iii) NBS, DAST; iv) Br$_2$

Figure 4.6

This again affords a new series of glycosyl donors which are activated under alternative conditions (see below), again extending the range of orthogonal donors and acceptors available for synthetic strategies.

Glycosyl halides

Glycosyl halides are often formed from glycosyl acetates, as described above, or from thioethyl pyranosides, via treatment with a source of the required halide ion. The incorporation of different halides into the anomeric position affords glycosyl donors with different reactivities and stabilities. The use of bromopyranosides has been widely exploited since these halopyranosides offer the most favourable combination of reactivity and stability. If glycosyl donors of enhanced reactivity are required, the bromides can be converted to the analogous iodides via a Finkelstein reaction, involving treatment of the glycosyl bromide with sodium iodide in acetone.

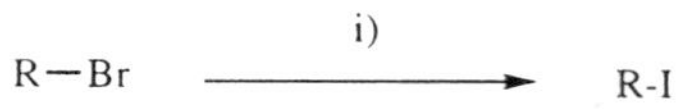

$$R-Br \xrightarrow{\text{i)}} R\text{-I}$$

i) NaI, acetone

Figure 4.7

In recent years glycosyl fluorides have also proved useful as glycosyl donors, even though they are considerably less reactive under normal conditions than other glycosyl halides. They are readily formed from thiopyranosides by treatment with *N*-bromosuccinimide (NBS) and diethylaminosulfur triflouride (DAST). Whilst glycosyl bromide donors are typically activated by treatment with a silver salt, such as silver silicate or silver triflate, glycosyl fluorides can be activated by treatment with the Lewis acid $BF_3.OEt_2$.

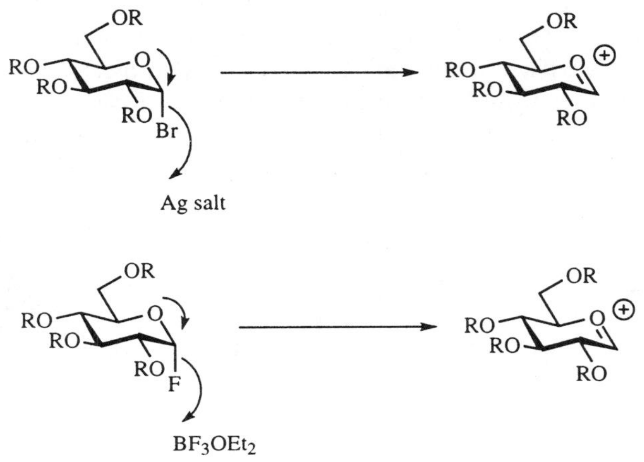

Glycosyl bromides can be activated with silver salts, whilst glycosyl fluorides can be activated with Lewis acids.

Figure 4.8

The former reaction, known as the Koenigs–Knorr reaction, has played, and indeed continues to play, a fundamental role in oligosaccharide assembly strategies.

Glycosyl selenides

Glycosyl bromides are also useful precursors to glycosyl selenides, which form a further useful group of glycosyl donors. They are again activated with *N*-iodosuccinimide and triflic acid (NIS/TfOH) under an identical procedure to that highlighted previously for thioglycosides (Figure 4.9).

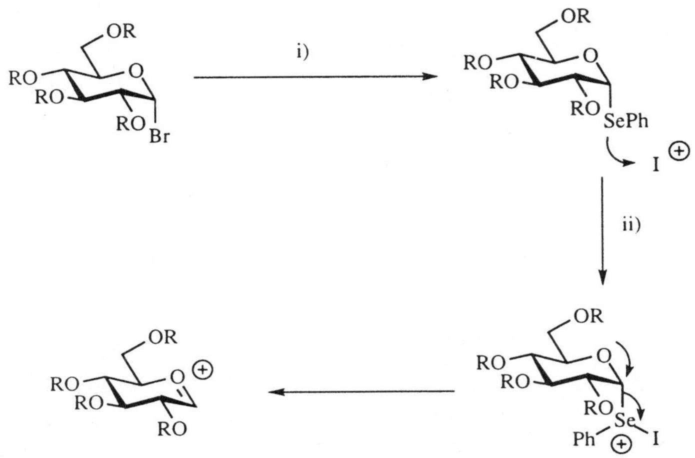

Glycosyl selenides are formed from glycosyl acetates by treatment with phenyl selenide. They are activated with NIS/TfOH.

i) PhSeSePh, NaBH$_4$; ii) NIS, TfOH

Figure 4.9

Trichloroacetimidates

The use of trichloroacetimidates as glycosyl donors is well documented in the literature. Trichloroacetimidate donors are easily formed via treatment of glycosyl alkoxides with trichloroacetonitrile and it is possible to control which anomer of the glycosyl donor forms by careful choice of the reaction conditions. In Figure 4.10 we have illustrated the mechanism for forming the trichloroacetimidate of a benzyl protected α-glucopyranoalkoxide. This would be achieved using NaH as base. If instead the reaction were performed in dichloromethane using potassium carbonate as base, entry to the analogous β-trichloroacetimidate would be achieved.

The different anomers of the trichloroacetimidates are formed by using different reaction conditions.

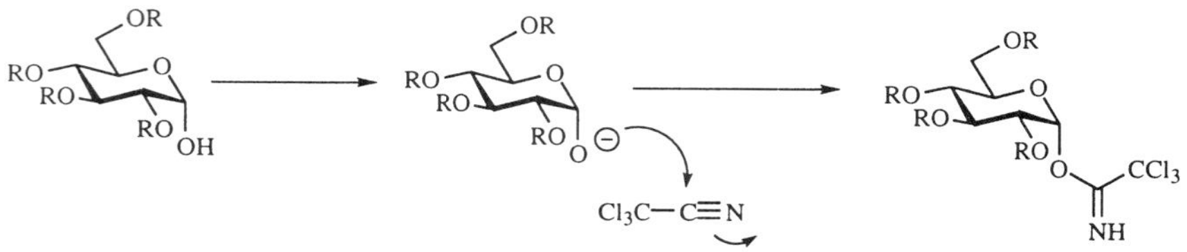

Figure 4.10

The β-alkoxide reacts quicker than the α-alkoxide so that the β-trichloroacetimidate is initially formed. This β-trichloroacetimidate is stable upon contact with the weak base potassium carbonate but unstable upon contact with the stronger base sodium hydride. Hence when potassium carbonate is used, the β-trichloroacetimidate is favoured whilst with sodium hydride epimerisation occurs to allow formation of the thermodynamically favoured α-trichloroacetimidate.

Activation of the trichloroacetimidate is achieved using a Lewis acid catalyst, for example $BF_3.OEt_2$, which again allows formation of the carbonium ion, which can be trapped with a carbohydrate acceptor, allowing entry to higher oligosaccharides.

Trichloroacetimidates are activated with Lewis acids.

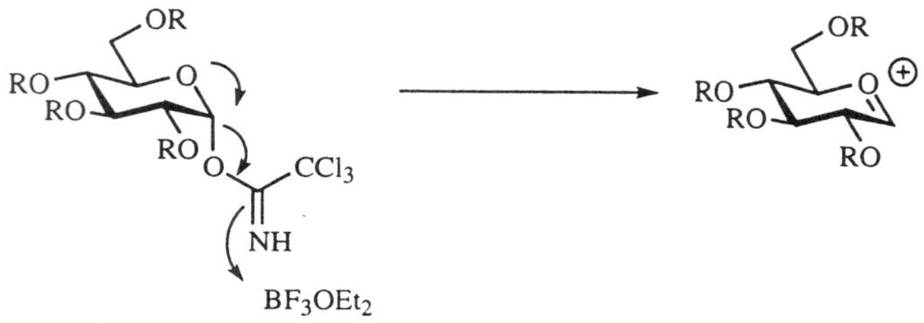

Figure 4.11

n-Pentenyl glycosides

O-Glycosides are not normally considered of use as carbohydrate donors. This is because alkoxide groups are poor leaving groups which are hard to displace under reaction conditions compatible with oligosaccharide synthesis. Indeed, in Chapter 3 we saw that many carbohydrate acceptors were themselves *O*-glycosides since they contained an anomeric group which was stable under the reaction conditions required for oligosaccharide synthesis. One exception to this rule is the use of *n*-pentenyl glycosides as carbohydrate donors. The *n*-pentenyl alkoxide group is stable under normal glycosidation conditions but can be converted to a good leaving group by treatment with *N*-bromosuccinimide (NBS) or more-reactive agents such as iodonium dicollidine perchlorate (IDCP), IOTf (generated *in situ* from NIS and TfOH) or TfOSiEt$_3$. The use of *n*-pentenyl glycosides as carbohydrate donors was pioneered by Fraser-Reid.

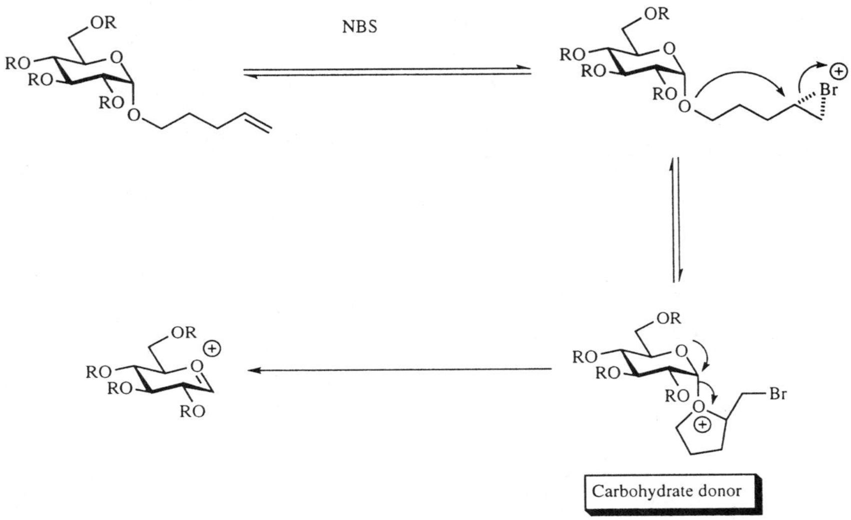

n-Pentenyl glycosides are activated by a source of halonium ion.

Figure 4.12

The electrophilic halonium ion initiates the activation process by reacting with the alkene moiety to afford a cationic intermediate. This is trapped in an intramolecular fashion by the anomeric oxygen atom to afford a cyclic oxocarbenium intermediate which is now a good carbohydrate donor molecule.

The *n*-pentenyl glycosides can be formed in a number of ways, including the direct glycosidation of unprotected sugars with 4-penten-1-ol under acidic conditions, glycosidation of glycosyl bromides under Koenigs–Knorr conditions, glycosidation of glycosyl acetates with 4-penten-1-ol under Lewis acid catalysis and rearrangement of ortho esters under acidic conditions (Figure 4.13).

Vinyl glycosides

One further class of anomeric alkoxides which are useful as glycosyl donors are the vinyl glycosides. Vinyl glycoside donors are generated *in situ* from the more stable 3-buten-2-yl glycosides by isomerisation with a rhodium catalyst (Figure 4.14). The donors can be activated to reaction with a Lewis acid and reactions are efficient with both primary and secondary sugar alcohols. The geometry of the newly formed glycosidic bonds depends on both the protecting groups present on the carbohydrate donor, and the reaction conditions used to initiate glycosidation.

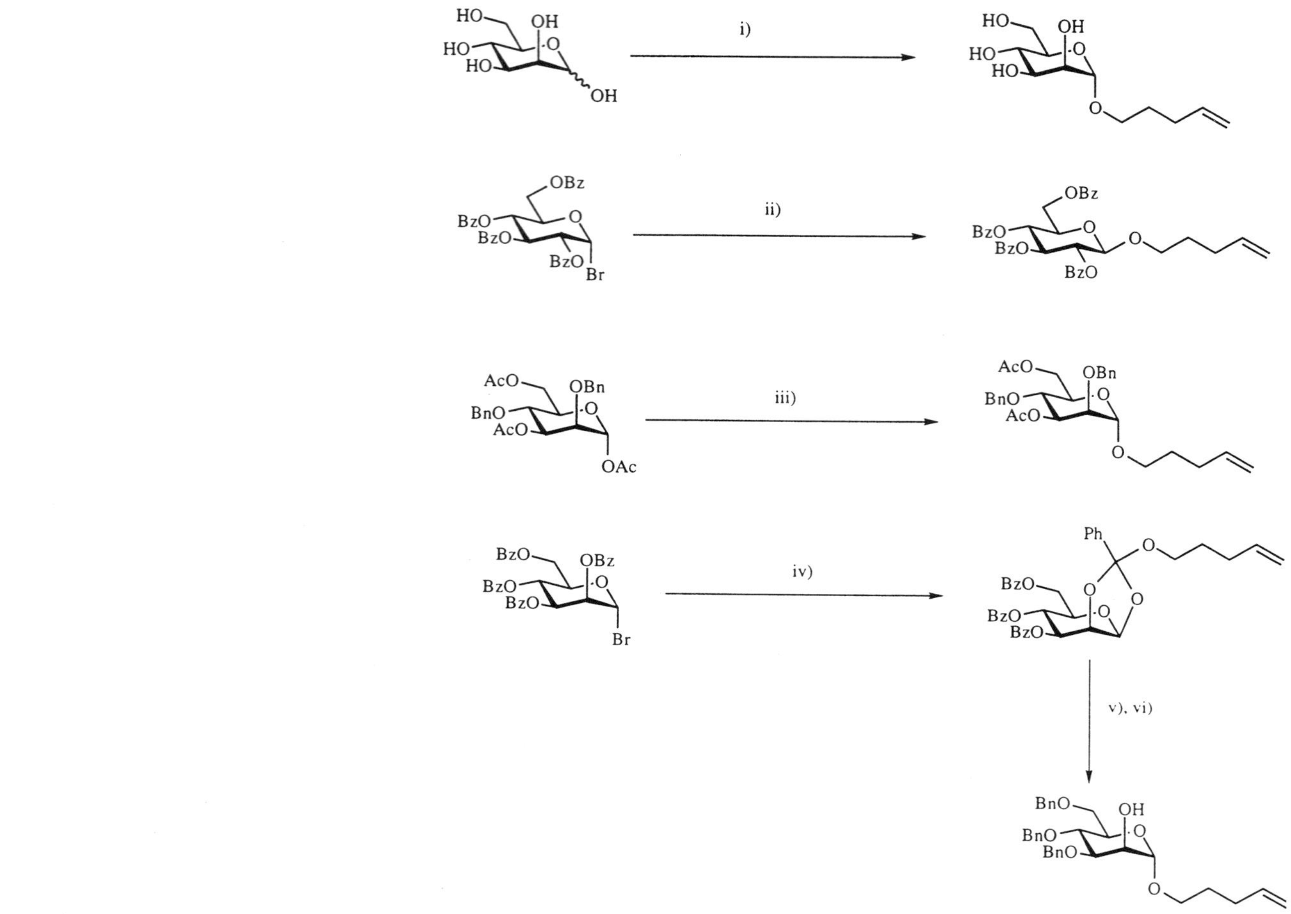

i) 4-Penten-1-ol, CSA; ii) 4-penten-1-ol, AgOTf; iii) 4-penten-1-ol, SnCl$_4$; iv) 4-penten-1-ol, lutidine; v) CSA; vi) NaOMe

Figure 4.13

Glycosyl sulfoxides

Thioglycosides are excellent precursors to glycosyl sulfoxides, the conversion being effected with oxidising agents such as *m*-chloroperbenzoic acid or Oxone®. Low temperatures are necessary to minimise further oxidation of the glycosyl sulfoxide to the glycosyl sulfone. The glycosyl sulfoxides act as donors in oligosaccharide assembly strategies upon treatment with triflic acid (TfOH) or triflic anhydride (Tf$_2$O) and an acid scavenger such as 2,6-di-*t*-butyl-4-methylpyridine.

It is believed that the reaction proceeds via initial reaction of the sulfoxide with triflic anhydride to generate the carbonium ion, which can react further in one of two ways. In the absence of other nucleophiles, the carbonium ion is trapped axially to afford a glycosyl triflate which can subsequently be attacked by acceptor molecules via an S$_N$2-like reaction. If, however, acceptors are present in the reaction medium at the time of carbonium ion formation, the carbonium ion is simply attacked at this stage without intermediate formation of the glycosyl triflate.

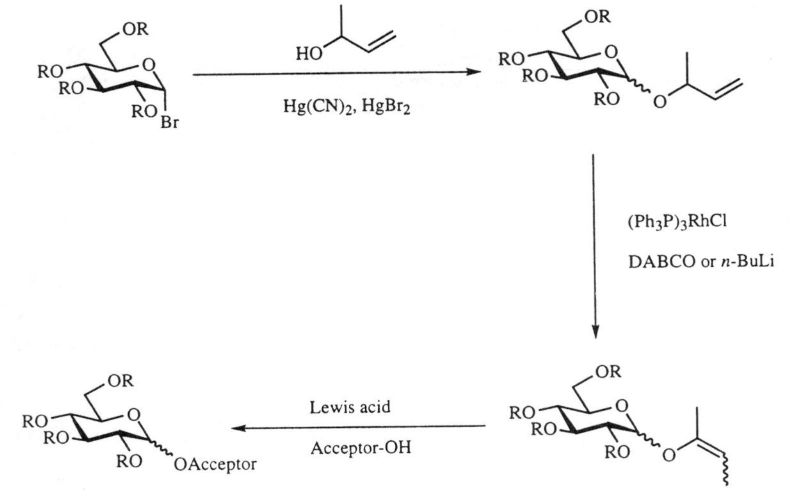

Vinyl glycosides are generated *in situ* from 3-buten-2-yl glycosides, and are activated by Lewis acids.

Figure 4.14

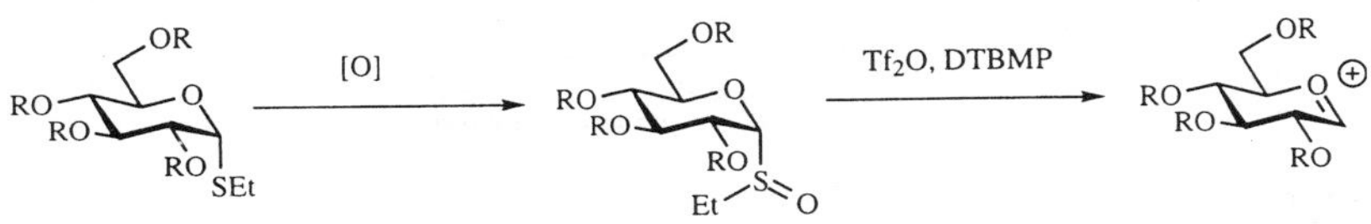

Glycosyl sulfoxides are activated with TfOH or Tf$_2$O and an acid scavenger.

Figure 4.15

Glycals

Glycals have been used extensively in carbohydrate-assembly strategies. A glycal is an unsaturated carbohydrate, and the class most useful as carbohydrate donors incorporate unsaturation between C-1 and C-2. A number of glycals are commercially available, but they are also readily available in the laboratory by elimination reactions. For example, glycosyl halides can readily be converted to the corresponding glycals in good yields under mild reaction conditions (Figure 4.17).

Danishefsky has published extensively on the use of glycals in oligosaccharide assembly strategies. In general, the glycal is activated to reaction by treatment with an epoxidising agent, typically dimethyldioxirane (DMDO), which forms the corresponding epoxide. This epoxide is not isolated but is treated *in situ* with the acceptor to allow entry to the extended unit, such as a disaccharide, which now contains a free hydroxyl group at C-2 (Figure 4.18).

The acceptor always approaches the epoxide intermediate from the top β-face, rather than the bottom α-face since the top face presents less steric hindrance.

In this chapter we have introduced a range of donor molecules that have proved their worth in carbohydrate-assembly strategies. In the next chapter we will consider the reaction of carbohydrate donors with carbohydrate acceptors to afford oligosaccharides.

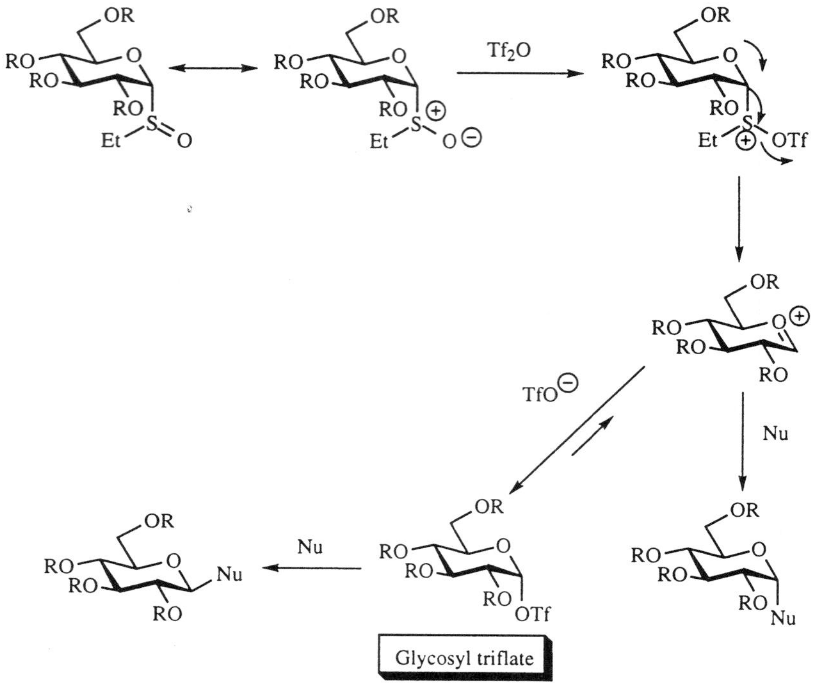

Figure 4.16

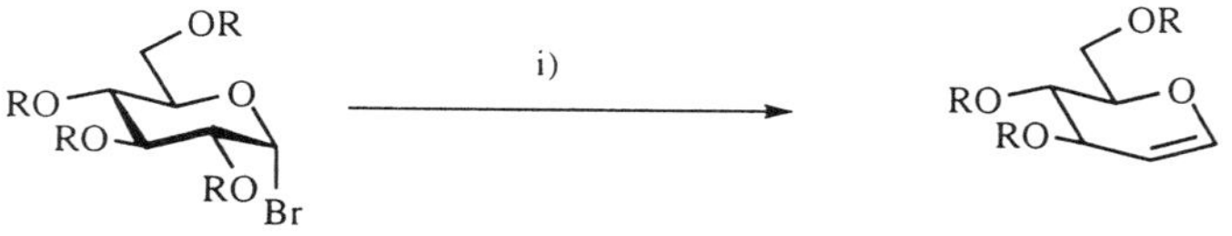

i) Zn, AcOH

Figure 4.17

Glycals are activated with DMDO.

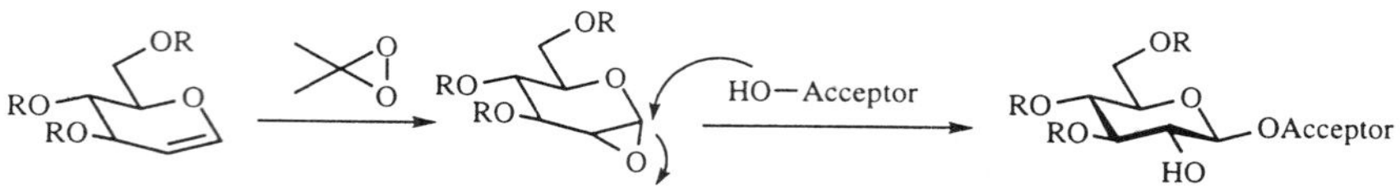

Figure 4.18

Further reading

1. Schmidt, R.R. (1986). New methods for the synthesis of glycosides and oligosaccharides. Are there alternatives to the Koenigs–Knorr method? *Angewandte Chemie International Edition in English*, **25**, 212–235.

2. Garegg, P.G. (1997). Thioglycosides as carbohydrate donors in organic synthesis. *Advances in Carbohydrate Chemistry and Biochemistry*, **52**, 179–206.

3. Witczak, Z.J., Czernecki, S. (1998). Synthetic applications of selenium-containing sugars. *Advances in Carbohydrate Chemistry and Biochemistry*, **53**, 143–200.

4. Schmidt, R.R., Kinzy, W. (1994). Anomeric-oxygen activation for glycoside synthesis: the trichloroacetimidate method. *Advances in Carbohydrate Chemistry and Biochemistry*, **50**, 21–124.

5. Fraser-Reid, B., Udodong, U.E., Ottoson, H., Merritt, J.R., Rao, C.S., Roberts, C., Madsen, R. (1992). *n*-Pentenyl glycosides in organic chemistry: a contemporary example of serendipity. *Synthetic Letters*, 927–942.

6. Boons, G-J., Isles, S. (1996). Vinyl glycosides in oligosaccharide synthesis 2. The use of allyl and vinyl glycosides in oligosaccharide synthesis. *Journal of Organic Chemistry*, **61**, 4262–4271.

7. Kahne, D., Walker, S., Cheng, Y., van Engen, D. (1989). Glycosylation of unreactive substrates. *Journal of the American Chemical Society*, **111**, 6881–6882.

8. Crich, D., Sun, S. (1997). Are glycosyl triflates intermediates in the sulfoxide glycosidation method? *Journal of the American Chemical Society*, **119**, 11217–11223.

9. Seeberger, P.H., Bilodeau, M.T., Danishefsky, S.J. (1997). Synthesis of biologically important oligosaccharides and other glycoconjugates by the glycal assembly method. *Aldrichimica Acta*, **30**, 75–92.

5 Assembly of oligosaccharides

Now that we have learnt how to synthesise specific donors and acceptors, we are in a position to consider the combination of these species to allow formation of oligosaccharides. A number of points must be considered for efficient synthesis of oligosaccharides. Suitable donor and acceptor molecules must be synthesised for each step of the reaction and care must be taken to ensure that any protecting groups incorporated within the molecules will remain intact until deprotection is required. Specific reagents are employed to activate the donor molecule and the reactive species thus formed is subsequently intercepted by the free hydroxyl group of an acceptor molecule, allowing formation of a glycosidic bond.

Synthetic methods must be available to allow synthesis of only the stereosiomer of choice.

We have already commented that different stereoisomers of biologically important compounds often exhibit different biological properties. If we require chemical access to a specific oligosaccharide to examine its biological properties in detail, it is of great importance to develop synthetic methods which allow access only to the stereosiomer of choice.

Furthermore, we have noted that the polyfunctional nature of carbohydrates can potentially allow entry to many different stereo- and regioisomers. Even if elaborate protecting-group strategies are employed to ensure that only the hydroxyl group of choice is involved in the formation of the new glycosidic bond, the possibility of forming two isomers at the new anomeric centre exists. We must devise ways to control which isomer results. Standard terminology has been introduced to distinguish the two stereoisomers that result, and these two stereoisomers are termed alpha- (α-) and beta- (β-) anomers. The two individual anomers are differentiated according to the relative orientation of the bond at the anomeric centre with that at the highest numbered asymmetric carbon of the unit. For example, consider methyl glucopyranoside. Two isomers exist, as highlighted in Figure 5.1. In the first case the anomeric bond lies in the same direction as the C-H band appended to C-5, whilst in the second isomer the anomeric bond lies in the opposite plane to the C-H band appended to C-5. By convention, the former isomer is termed the α-anomer and the latter the β-anomer.

The development of new synthetic methods to allow entry to specific anomers has received considerable attention throughout the advancement of carbohydrate chemistry. Some of the methods available for governing the formation of these specific anomers are highlighted below.

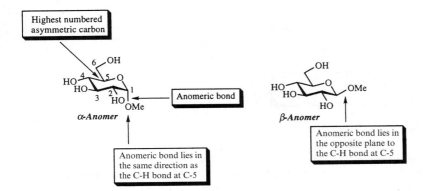

The two isomers which result from the change in configuration at C-1 are termed anomers.

Figure 5.1

Methods to control the configuration of newly formed anomeric bonds

Via the anomeric effect

In the absence of any external factors, newly formed glycosidic bonds will prefer to exist in the α-configuration rather than the β-configuration, owing to the anomeric effect. The anomeric effect is a stereoelectronic effect which seeks to explain why electronegative substituents in the anomeric position of pyranose rings tend to exist in an axial orientation rather than an equatorial position. This is, of course, opposite to the preferred conformation for cyclohexane rings, where bulky substituents prefer to occupy equatorial positions to minimise steric interactions.

Electronegative substituents in the anomeric position of pyranose rings tend to exist in an axial orientation rather than an equatorial position.

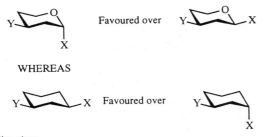

X = Electronegative atom

Figure 5.2

There are many hypotheses which explain the anomeric effect, but, in simple terms, it is due to cyclic acetal systems preferring their substituents to display a gauche arrangement around two contiguous C–X bonds (where X = electronegative element). For carbohydrate systems, one of the C–X bonds is a C–O bond. When the two bonds display the gauche arrangement, the lone pair of electrons on the oxygen atom within the ring lies antiperiplanar to the substituent C–X bond. This is stereoelectronically favoured, as donation of the lone pair of electrons on the ring oxygen into a vacant σ* orbital of the adjacent hetereoatom can occur (Figure 5.3).

Any reaction which is performed under thermodynamic conditions will therefore favour formation of the α-anomer, although this effect

Favoured over

Donation of electrons
to empty σ* orbital

Figure 5.3

decreases as the polarity of the solvent increases. This means that if entry to the β-anomer is required, other specific strategies are required. Some suitable methodologies for synthesising β-anomers are discussed below.

Via careful selection of solvent

The effect of solvent upon the stereoselectivity of glycosidation reactions is an important factor which should not be ignored when designing oligosaccharide synthesis. Often specific solvents can be used that allow entry to the anomer which would be disfavoured under alternative conditions. For example, although donors displaying non-ester groups at C-2 generally favour formation of α-anomers, due to the anomeric effect, the less thermodynamically favoured β-anomer can be accessed if the reaction is conducted in a 'participating solvent' such as acetonitrile. This is because acetonitrile is itself nucleophilic and is able to intercept the carbonium ion formed upon activation of the donor, which initially allows access to the α-nitrilium intermediate. When a second nucleophile is introduced, such as a carbohydrate acceptor, an S_N2 reaction occurs with the resultant inversion reaction affording the β-anomer.

'Participating solvents' such as MeCN can allow entry to β-linked glucopyranoside oligosaccharides if non-ester groups are incorporated at C-2.

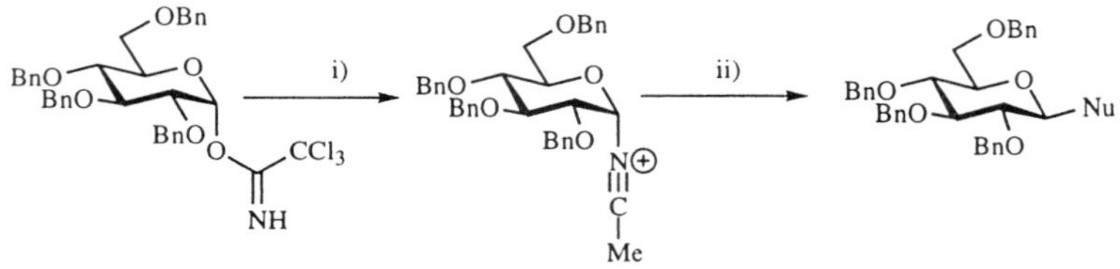

i) MeCN/ ii) Nucleophile

Figure 5.4

Via neighbouring group participation

One method for forming β-glucopyranoside oligosaccharides is to incorporate an ester protecting group at C-2. Initially, activation of the donor results in the formation of a carbonium ion, via loss of the anomeric group at C-1. If an ester group is available at C-2 this can interact with the carbonium ion in an intramolecular fashion as shown for a glucopyranoside derivative in Figure 5.5. This stabilising interaction is termed 'neighbouring group participation (NGP)' or 'anchimeric assistance'. The new intermediate which is produced is now attacked in an intermolecular fashion by the acceptor molecule allowing access to the target molecule. Although two stereoisomers could result from this reaction, for the glucopyranoside derivative depicted, the β-anomer is favoured. This is because the donor will encounter less steric hindrance if it approaches the anomeric centre from above the plane rather than from below, allowing entry to the β-linked stereoisomer.

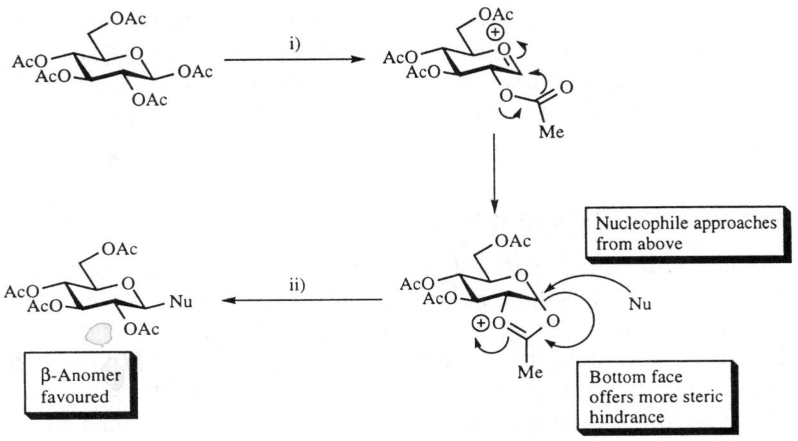

NGP allows entry to β-linked glucopyranoside oligosaccharides if an ester group is present at C-2.

Nucleophile approaches from above

Bottom face offers more steric hindrance

β-Anomer favoured

i) BF$_3$.OEt$_2$; ii) Nucleophile

Figure 5.5

The orientation of the C-2 ester group will obviously effect the overall stereochemical outcome of the reaction. If the C-2 group is equatorial, as is the case for gluco- and galactopyranoside derivatives, the β-isomer will result. However, for mannopyranoside derivatives, in which the C-2 group is axial, the donor will approach from below the plane, allowing access to the α-anomer (Figure 5.6).

Thus whilst the anomeric effect favours formation of α-linked products

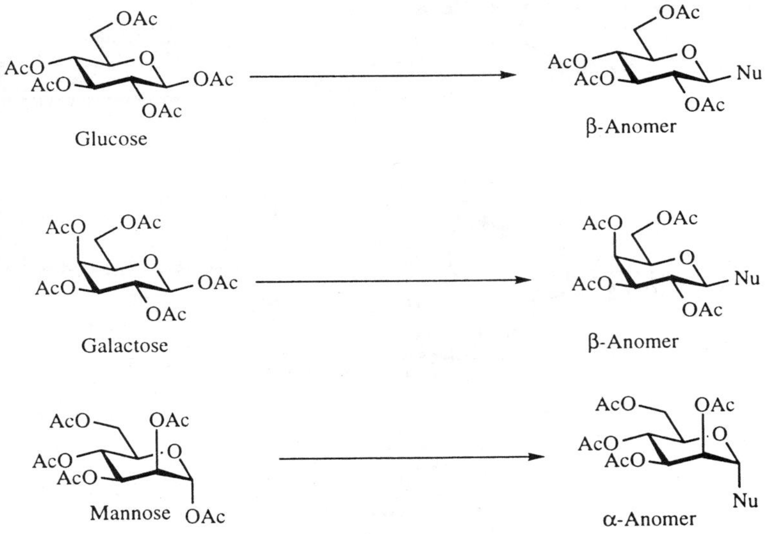

Glucose

β-Anomer

Galactose

β-Anomer

Mannose

α-Anomer

Figure 5.6

with gluco-, galacto- and mannopyranosides, incorporation of an ester group at C-2 can allow access to the complimentary β-linked stereoisomer for gluco- and galactopyranosides. This alternative strategy is not, however, suitable for entry to β-linked mannopyranoside derivatives since, as we have seen, neighbouring group participation also allows access to α-linked mannopyranosides. Even if non-participating protecting groups are incorporated at C-2, the synthesis of β-linked mannopyranosides remains particularly difficult, and a range of strategies specifically applicable to the synthesis of these biologically important linkages has been developed.

β-Linked mannopyranosides are difficult to obtain using standard glycosidation strategies.

Synthesis of β-linked mannopyranosides

A whole multitude of naturally occurring oligosaccharides contain β-linked mannopyranosides, emphasising the requirement for developing synthetic methods for entry to these structures. One specific oligosaccharide of this type, alongside the core structure of *N*-linked glycoproteins which contain β-mannopyranosides, is highlighted in Figure 5.7.

Some of the synthetic strategies which have been developed for the synthesis of β-mannopyranosides are discussed in detail below. Since β-linked glucopyranoside derivatives are easily synthesised using neighbouring group participation, several syntheses of β-linked mannopyranosides employ β-linked glucopyranosides as key starting materials. This is because once the β-glycosidic linkage has been formed, inversion of the hydroxyl

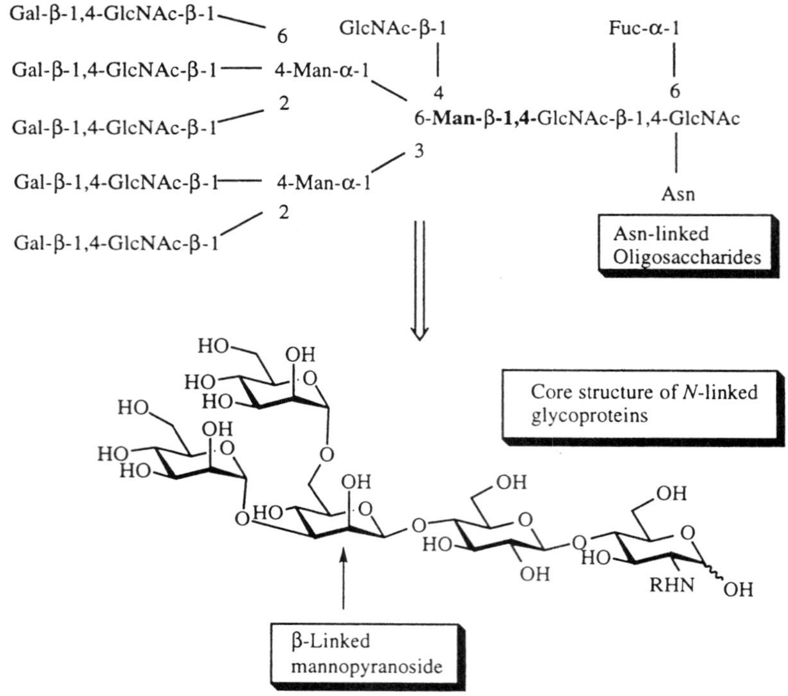

Figure 5.7

group at C-2 is all that is necessary to allow formation of the β-mannosidic linkage. The application of this idea to two synthetic strategies is described below.

Using silver catalysts

We have seen in Chapter 4 that glycosyl halides are useful donors for the synthesis of oligosaccharides. In particular, glycosyl bromides have found widespread use in strategies such as the Koenigs–Knorr glycosidation methodology, where the donors are usually activated for reaction with silver promoters. It has been discovered that when bulky silver promoters are used, for example silver silicate, high levels of stereocontrol can be effected in the reactions that follow. This is because when the bulky promoter binds to the anomeric group, the bottom face of the donor becomes effectively blocked to approach by the nucleophile. The nucleophile will therefore prefer to approach the donor from the top face, affording a preponderance of the β-anomer. This approach is often used for the synthesis of β-linked mannopyranosides by employing 2-oxoglycosyl bromides as donors. The 2-oxoglycosyl bromides are themselves prepared from glucopyranosides. Glycosidation of these donors affords β-linked glycosidulose derivatives which, after stereoselective reduction with sodium borohydride, preferentially affords β-linked mannopyranosides rather than β-linked glucopyranosides.

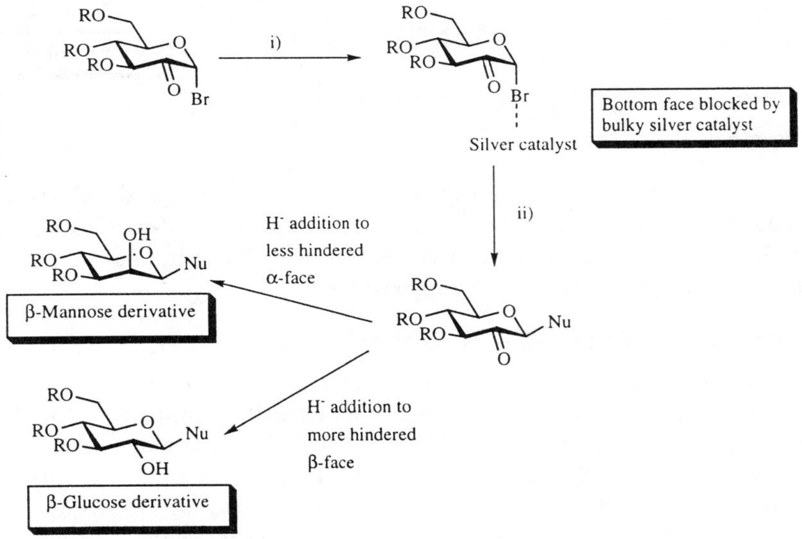

i) Silver silicate; ii) Nucleophile

Figure 5.8

β-Linked mannopyranosides can be obtained from glucosyl bromides if silver catalysts are used in the glycosidation reaction.

This approach has been widely utilised by Lichtenthaler for the synthesis of β-linked mannose-containing *N*-linked core oligosaccharides, as well as for the synthesis of β-linked mannosamine units. Although the mannosamine units are not as widespread in nature as β-linked mannose units, they are key constituents of a range of naturally occurring oligosaccharides. For example, they are found in the repeating units of a range of capsular polysaccharides displayed by *Streptococcus pneumoniae* bacteria.

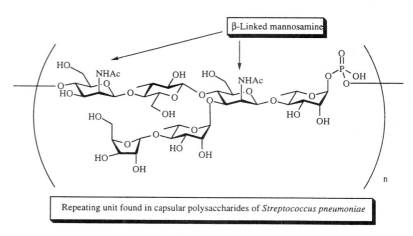

Figure 5.9

An analogous synthetic procedure to that described above can also be used for the synthesis of β-linked mannosamine units.

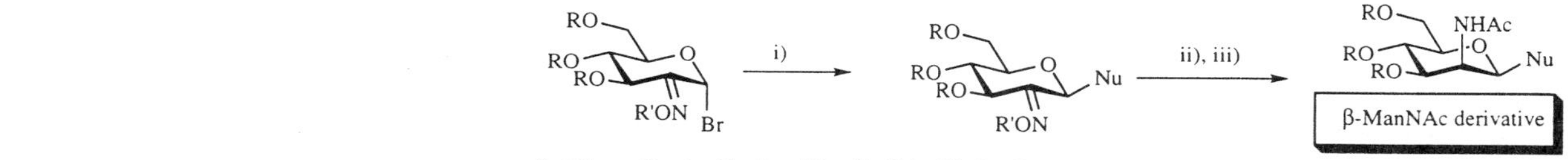

i) Silver silicate, Nucleophile; ii) BH_3; iii) Ac_2O, pyridine

Figure 5.10

After silver silicate promoted glycosidation, the β-linked oximes are reduced in a stereoselective fashion to allow entry to the β-mannosamine-containing oligosaccharides, in preference to β-glucosamine-containing oligosaccharides, using borane in THF.

By inversion of the C-2 stereocentre

An alternative approach for the synthesis of β-mannopyranosides involves S_N2 displacement of an excellent leaving group from the C-2 position of glucopyranosides. Since the S_N2 displacement occurs with inversion of stereochemistry, the equatorial group at C-2 of the glucopyranoside is converted into an axial C-2 group of a mannopyranoside. For example, if the hydroxyl group at C-2 of glucopyranosides is converted into a 2–*O*-(*N*-imidazolylsulfonyl) group, it can be displaced by a benzoate ion to afford a β-linked mannose unit with a benzoate group at C-2. This can be deprotected under routine conditions to afford the β-linked mannose units.

This approach is also useful for the synthesis of β-linked mannosamine structures. Analogous S_N2 displacement of the leaving group at C-2 with the azide anion is utilised to allow introduction of a nitrogen substituent at this position. Subsequent reduction of the azide functional group can then be used to afford the β-linked mannosamine structures.

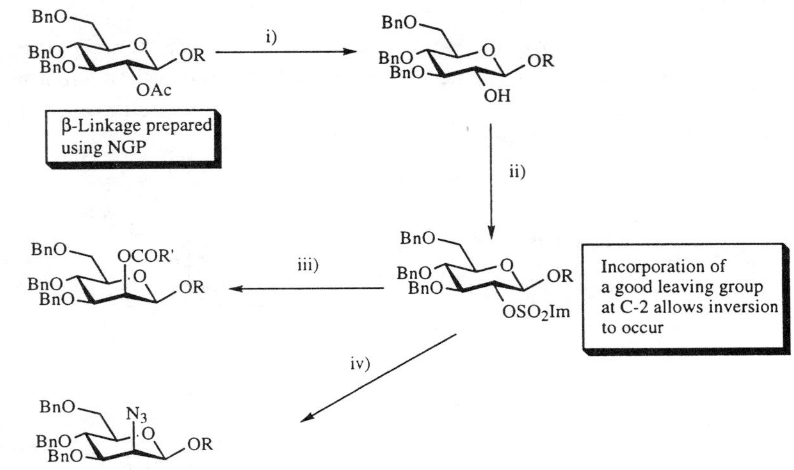

i) NaOMe, MeOH; ii) $ImSO_2Cl$; iii) NaOBz; iv) NaN_3

Figure 5.11

S_N2 displacement of leaving groups at C-2 of β-linked glucopyranosides allows entry to β-linked mannopyranosides.

*Inter*molecular nucleophilic substitutions of this type are not always of sufficient efficiency to allow access to β-mannose units in acceptable yields. When the yield of this reaction limits the efficiency of a synthetic strategy, an *intra*molecular displacement reaction can be employed. For example, consider a glucopyranoside derivative which incorporates a carbamate-protecting group at C-3 and a good leaving group, such as a triflate at C-2.

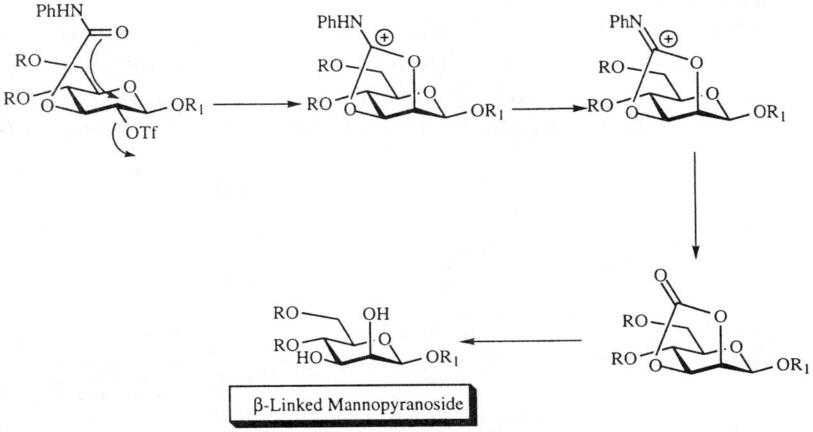

Figure 5.12

Once the carbamate group is deprotonated, intramolecular substitution by the newly formed nucleophile can occur at C-2 . The resultant intermediate is prone to hydrolysis to form a carbonate which, under acidic conditions, undergoes deprotection to afford a β-linked mannopyranoside with free hydroxyl groups at both C-2 and C-3.

Using donors tethered to acceptors

If a donor and acceptor could be tethered together in the correct orientation to allow addition of the acceptor to the top β-face of an activated donor, entry to β-linked mannopyranosides could be achieved. This general principle is illustrated in Figure 5.13.

Tethered donors and acceptors allow entry to β-linked mannopyranosides.

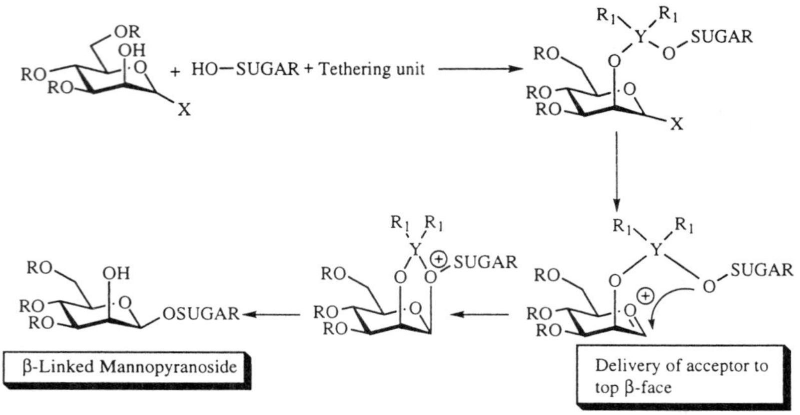

Figure 5.13

In reality, this has proved possible using a range of tethering agents. For example, Stork and Hindsgaul have illustrated that silyl and dimethyl acetal tethers are suitable for this strategy. In both cases the carbohydrate acceptor can be linked to the carbohydrate donor either via its most reactive C-6 hydroxyl group or via the less reactive secondary hydroxyl groups at C-2, C-3 and C-4.

Carbon-tethered species have also been utilised in a similar strategy by Ogawa. In this case, the C-2 hydroxyl group is converted into its *p*-methoxybenzyl ether. Since the benzylic position of *p*-methoxybenzyl ethers can be activated using oxidising agents such as 2,3-dichloro-5,6-

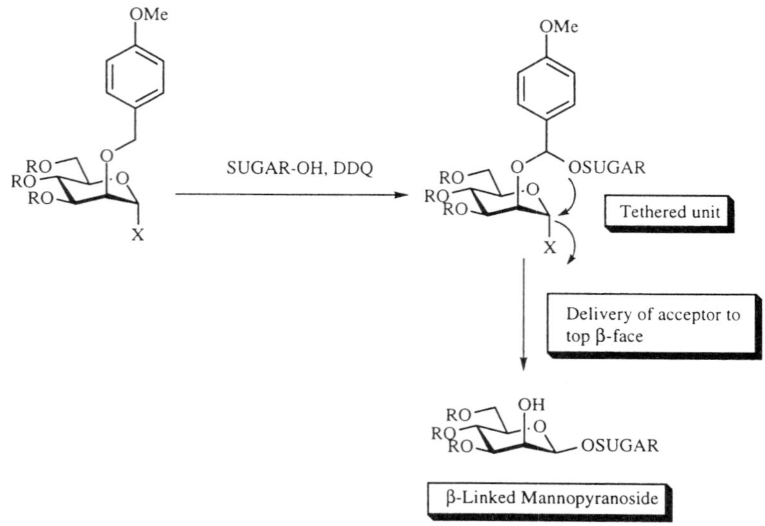

Figure 5.14

dicyano-1,4-benzoquinone (DDQ), carbocation formation can be initiated at the benzylic position of the C-2 ether. Once the carbocation is formed it is trapped with a sugar acceptor to form a carbon-tethered molecule. Activation of the donor component of this molecule and subsequent interception of the carbocation thus formed with the tethered acceptor then allows entry, after hydrolysis of the tether, to the desired β-linked mannopyranosides.

We have now seen that there are a number of ways in which we can synthesise specific anomers of carbohydrates. Even after we have decided which strategy to use, there are many ways to actually approach the synthesis of our target oligosaccharides. For example, in the literature we find that more traditional strategies rely upon stepwise assembly of the oligosaccharide. These normally involve activation of a donor, trapping of the resultant reactive intermediate with an acceptor and purification of the resultant saccharide, prior to performing further glycosidation reactions on the newly formed molecule. More modern strategies seek to synthesise molecules in one-pot which means that saccharides which are synthesised *en-route* to larger oligosaccharides are not isolated, but reacted *in situ*. This has the advantage of minimising purification and isolation procedures. This latter approach is more efficient, so long as the coupling reaction of the donor with the acceptor proceeds efficiently in each case, and produces the required oligosaccharide in excellent yield, and with exceptional stereocontrol. A number of strategies have recently been developed to allow one-pot synthesis of oligosaccharides. In every case, the methods concentrate upon selective activation of one donor in the presence of another potential donor. This can be done in a number of ways, including the use of orthogonal anomeric groups, and by tuning of reactivity.

One-pot synthesis of oligosaccharides

Via orthogonal glycosidation strategies

Orthogonal glycosidation strategies have allowed one-pot entry to target oligosaccharides. In such approaches, two donors displaying *different* anomeric groups (X and Y) which are activated by *different* promoters are employed in the general strategy outlined below in Figure 5.15.

Two different anomeric groups are incorporated within the two donors and, importantly, each anomeric group survives the reaction conditions necessary to activate the other donor molecule. For example, if a mixture of a thiopyranoside and a fluoropyranoside were treated with the promoter system *N*-iodosuccinimide and silver triflate (NIS, AgOTf), only the thiopyranoside would be activated to allow formation of its analogous carbonium ion. Therefore, if the fluoropyranoside displayed a free hydroxyl group which could trap this carbonium ion, then entry to a fluoropyranoside disaccharide would occur. Subsequent *in situ* treatment of this new disaccharide donor with the Lewis acid $BF_3.OEt_2$ and an acceptor whose anomeric group remained intact on contact with this activating system would then allow entry to a trisaccharide donor. The above process can be repeated indefinitely, allowing entry to large oligosaccharides. The product saccharides must, however, form in excellent yields

Orthogonal glycosidation strategies employ donors that display anomeric groups which are activated by different promoters.

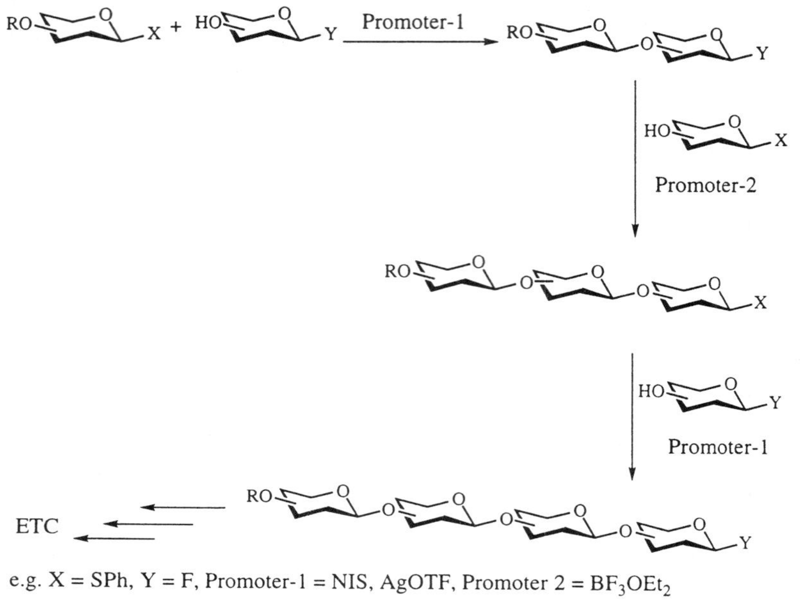

e.g. X = SPh, Y = F, Promoter-1 = NIS, AgOTF, Promoter 2 = BF$_3$OEt$_2$

Figure 5.15

with excellent stereoselectivity if the reaction is to be performed in one pot without the need for purification or isolation of the intermediates.

Via tuning of reactivity

A further concept which potentially allows one-pot synthesis of oligosaccharides involves controlling the relative reactivity of glycosyl donors. For example, consider two donors, such as the *n*-pentenyl glycosides displayed in Figure 5.16, both of which can be activated by one promoting agent.

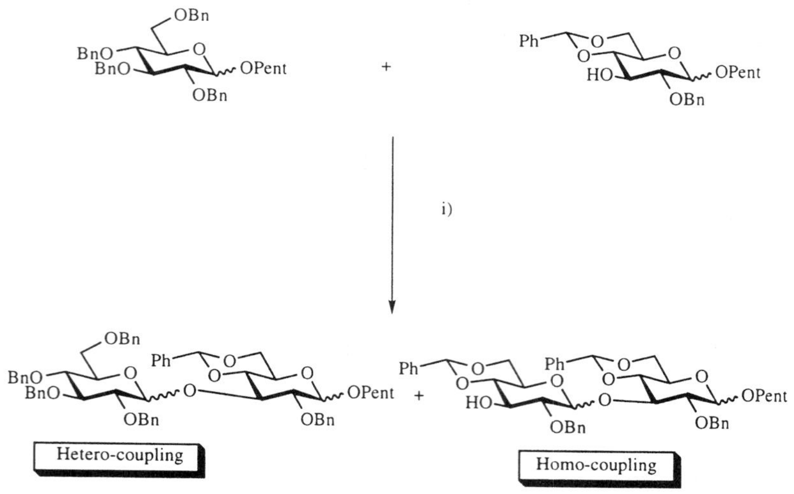

i) NBS

Figure 5.16

If both donors were activated by the promoting agent to the same extent, then access to both the disaccharide resulting from hetero-coupling and the disaccharide resulting from homo-coupling would be possible. However, if one donor were designed to incorporate structural features which diminished its ability to be activated by the promoter, then on addition of a limiting amount of the promoting agent only the more reactive donor would form a carbonium ion. Entry into only one disaccharide would then be achieved with no homo-coupling occurring.

In theory, a whole series of donors with differing abilities to be activated can be developed. This means that strategies can be designed which employ molecules which can serve both as donors and acceptors. Such molecules which can act as either donors or acceptors without the need for any functional-group interconversions are termed 'hermaphrodites'. Upon addition of the promoting agent, only the most reactive donor will be activated, and this will be trapped by the hermaphrodite which, at this stage, acts as an acceptor. Such observations have led Fraser-Reid and others to design syntheses based on 'armed' and 'disarmed' reaction components. The term 'armed' is allocated to the reaction component with the greater ability to be activated by the promoting agent. Thus the 'armed' donor is able to form a carbonium ion when limiting quantities of activating agent are employed, and the carbonium ion is trapped by the 'disarmed' acceptor.

In practise, many structural features can be incorporated within the donors and hermaphrodites to alter their ability to be activated by promoting agents. For example, it has been demonstrated that different protecting groups can profoundly alter the reactivity of donors. Returning to the example introduced in Figure 5.16, if limiting quantities of the activating agent are utilised, only the fully benzylated *n*-pentenyl glycoside is activated, so that entry to only the disaccharide resulting from hetero-coupling results. This is because formation of the carbonium ion, by loss of the anomeric substituent, results in flattening of the pyranose chair. The rigidity of the benzylidene acetal incorporated within the alternative donor resists flattening of the ring, thus reducing the ability of this donor to be activated. The newly formed disaccharide derived from this hetero-coupling process is open to subsequent activation with a further equivalent of promoting agent, and could be trapped with a further equivalent of hermaphrodite acceptor, without the need for isolation and purification of the intermediate disaccharide, to afford larger saccharides. Such strategies have allowed synthesis of large oligosaccharides in one pot.

In theory, a whole range of protecting groups can be incorporated within donors and hermaphrodites to provide a series of donors with varying reactivity. For example, donors displaying ester-protecting groups are 'disarmed' to activation compared with similar donors bearing ether-protecting groups.

In particular, the C-2-protecting group has the greatest influence on the tuning of reactivity, presumably due to its proximity to the anomeric centre. For the example illustrated in Figure 5.17, little or no homo-coupled disaccharide results when the 'armed' donor is trapped by the 'disarmed' hermaphrodite. Moreover, the resulting disaccharide displays a thioethyl group at the anomeric position, meaning that further activation

The reactivity of molecules towards glycosidation can be controlled by incorporating specific structural components within the molecule.

An 'armed' donor is able to form a carbonium ion when limiting quantities of activating agent are employed, and the carbonium ion is trapped by the 'disarmed' acceptor.

A benzylidene acetal can 'disarm'

Esters 'disarm' donors.

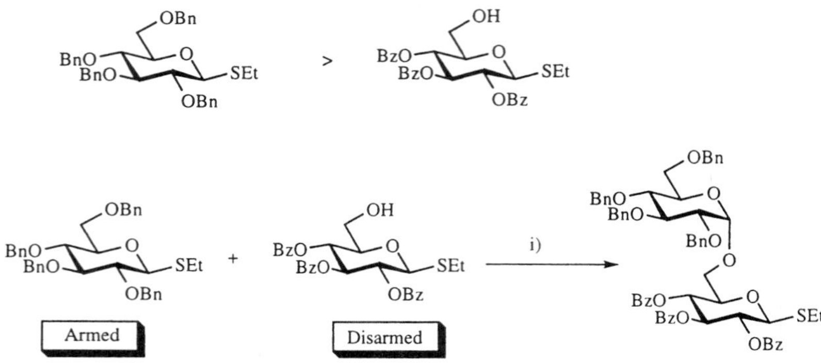

i) NIS, TfOH

Figure 5.17

and reaction with an acceptor moiety could be utilised to allow entry to larger oligosaccharides.

Further protecting groups which are also of value for mediating reactivity include the dione-protecting groups previously described for the protection of *trans*-hydroxyl groups. It has been found that incorporation of a dione-protecting group diminishes the donor's ability to form a carbonium ion. Again, this is a result of carbonium ion formation requiring flattening of the chair conformation, which is disfavoured by the dione-protected compounds due to the rigidity imparted on the ring system. This strategy has been employed by Ley and co-workers in a series of syntheses of large oligosaccharides. Figure 5.18 illustrates one early example which has been utilised for the synthesis of a rhamnose trisaccharide.

Entry to the disaccharide intermediate is achieved by reaction of an 'armed' thioethyl donor with a 'disarmed' thioethyl hermaphrodite. The intermediate disaccharide formed can be activated with a further equivalent of the promoting system and trapped with a further 'disarmed' acceptor to afford the target trisaccharide. The second acceptor moiety is 'disarmed' both by virtue of the diacetal-protecting group and by incorporating a less-reactive anomeric substituent. Interestingly, this reaction was performed both in a one-pot synthesis and via a stepwise approach, but better yields were afforded when the disaccharide intermediate was *not* isolated, i.e. better yields were obtained under the one-pot strategy.

The notion of tuning reactivity via incorporation of different leaving groups at the anomeric position of the donor has been studied in depth. For example, thiopyranosides and selenopyranosides can both be activated by the NIS/TfOH promoter system, but selenopyranosides are more prone to activation than thiopyranosides (Figure 5.19).

Thus, if a thioethyl hermaphrodite molecule is employed in conjunction with a selenopyranoside, activation with one equivalent of NIS/TfOH will afford the thioethyl disaccharide, which is again a useful intermediate for further glycosidation reactions. It has also proved possible to alter the reactivity of glycosyl donors by varying the size of the anomeric substituent. When cyclohexane groups are incorporated at the anomeric centre, donors with reactivity intermediate between that of the fully benzylated and fully benzoylated counterparts is achieved (Figure 5.20).

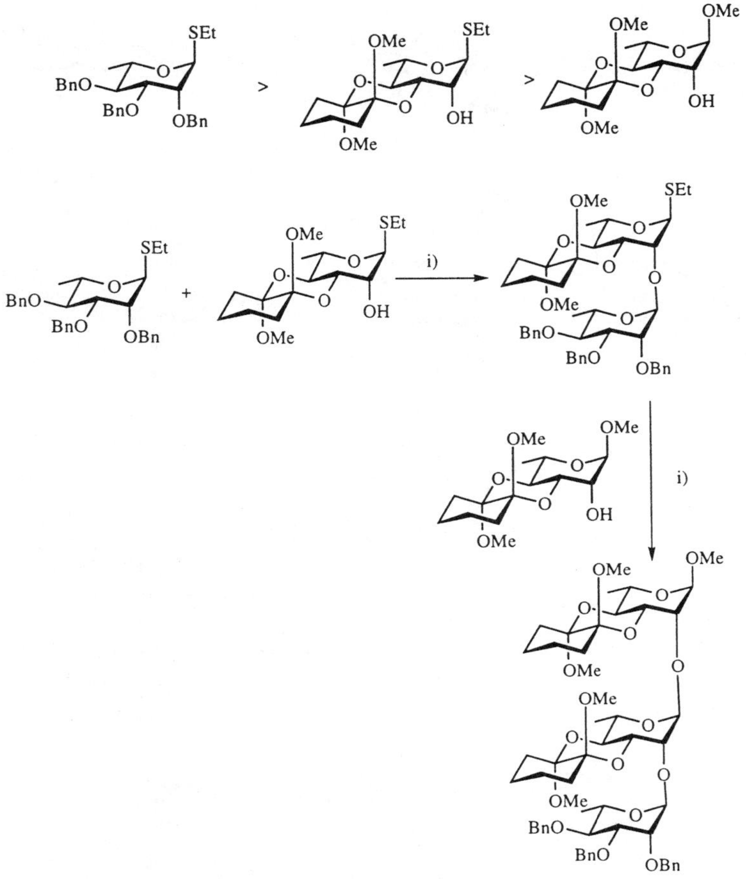

The dione-protecting group for *trans*-hydroxyl groups 'disarm' donors.

i) NIS, TfOH

Figure 5.18

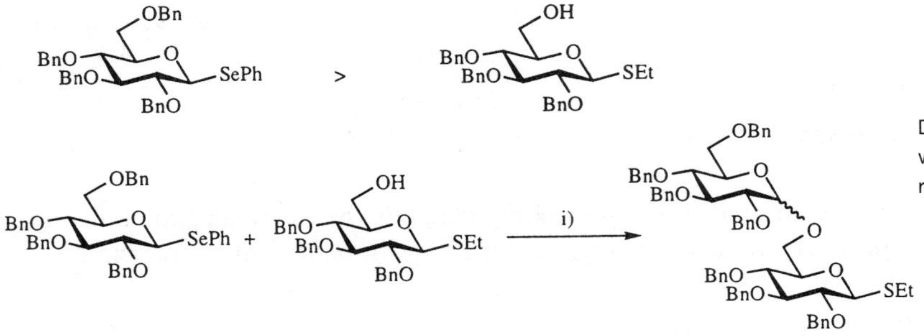

Different anomeric groups can be chosen which provide glycosyl donors of varying reactivity.

i) NIS, TfOH

Figure 5.19

In this chapter we have introduced some methods for controlling the geometry of newly formed glycosidic bonds, as well as some specific strategies which can be employed for efficient syntheses of oligosaccharides. In Chapter 6 we will illustrate these methods, as well as some of the tactics

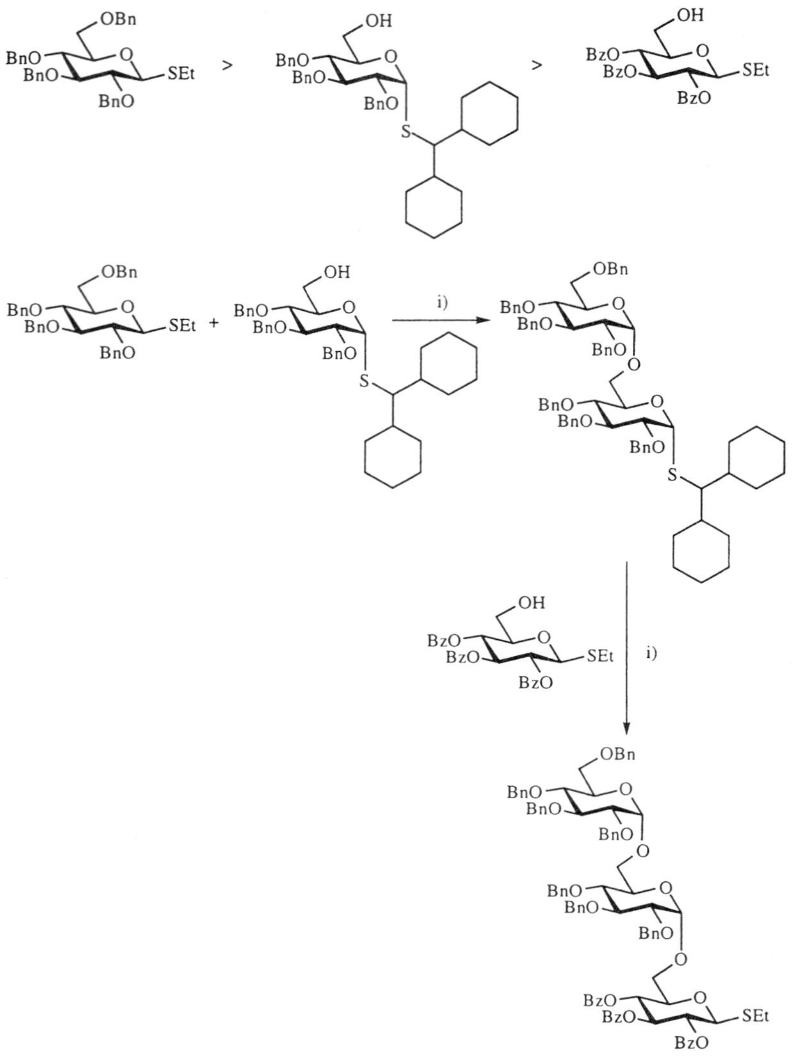

i) NIS, TfOH

Figure 5.20

we have covered in the preceeding chapters, by discussing some literature reports of syntheses of oligosaccharides with biological importance.

Further reading

1. Boons, G-J. (1996). Strategies in oligosaccharide synthesis. *Tetrahedron*, **52**, 1095–1121.

2. Boons, G-J. (1996). Recent developments in chemical oligosaccharide synthesis. *Contemporary Organic Synthesis*, **3**, 173–200.

3. Kaji, E., Lichtenthaler, F.W. (1993). 2–Oxo- and 2–oximinoglycosyl halides: versatile glycosyl donors for the construction of β-D-mannose

and β-D-mannosamine-containing oligosaccharides. *Trends in Glyco-science and Glycotechnology*, **5**, 121–142.

4. Günther, W., Kunz, H. (1992). Synthesis of β-D-mannosides from β-D-glucosides *via* an intramolecular S_N2 reaction at C-2. *Carbohydrate Research*, **228**, 217–241.

5. Barresi, F., Hindsgaul, O. (1994). The synthesis of beta-mannopyrano-sides by intramolecular aglycon delivery—scope and limitations of the existing methodology. *Canadian Journal of Chemistry*, **72**, 1447–1465.

6. Crich, D., Sun, S. (1997). Direct synthesis of β-mannopyranosides by the sulfoxide method. *Journal of Organic Chemistry*, **62**, 1198–1199.

7. Kanie, O., Ito, Y., Ogawa, T. (1994). Orthogonal glycosylation strategies in oligosaccharide synthesis. *Journal of the American Chemical Society*, **116**, 12073–12074.

8. Fraser-Reid, B., Wu, Z.F., Udodong, U.E., Ottoson, H. (1990). Armed-disarmed effects in glycosyl donors—rationalization and side tracking. *Journal of Organic Chemistry*, **55**, 6068–6070.

9. Ley, S.V., Priepke, H.W.M. (1994). Cyclohexane-1,2-diacetals in synthesis 2. A facile one-pot synthesis of a trisaccharide unit from the common polysaccharide antigen of group-B *Streptococci* using cyclohexane-1,2-diacetal protected rhamnosides. *Angewandte Chemie International Edition in English*, **33**, 2292–2294.

10. Ziegler, T. (1994). The selective blocking of trans-diequatorial, vicinal diols; application in the synthesis of chiral building blocks and complex sugars. *Angewandte Chemie International Edition in English*, **33**, 2272–2275.

11. Green, L., Hinzen, B., Ince, S.J., Langer, P., Ley, S.V., Warriner, S.L. (1998). One-pot synthesis of penta- and hepta-saccharides from monomeric mannose building blocks using the principles of orthogonality and reactivity tuning. *Synthetic Letters*, 440–442.

12. Boons, G-J., Grice, P., Leslie, R., Ley, S.V., Yeung, L.L. (1993). Dispiroketals in synthesis 5. A new opportunity for organic synthesis using differentially activated glycosyl donors and acceptors. *Tetrahedron Letters*, **34**, 8523–8526.

13. Boons, G-J., Geursten, R., Holmes, D. (1995). Chemoselective glycosidations 1. Difference in size of anomeric leaving group can be exploited in chemoselective glycosylations. *Tetrahedron Letters*, **36**, 6325–6328.

6 Literature examples of oligosaccharide synthesis

In this chapter we will describe some recent examples of oligosaccharide synthesis. We hope that these syntheses will further exemplify the methods and procedures introduced in the previous chapters, and also illustrate that you have now met sufficient reactions to devise syntheses of even complex carbohydrates. Literature references to the original articles reporting the cited work are quoted for further reference at the end of the chapter.

An efficient synthesis of a high-mannose-type nonasaccharide

For our first example we have chosen the total synthesis of a high-mannose-type nonasaccharide which is part of the gp120 glycoprotein of the viral coat of HIV-1. A range of high-mannose oligosaccharides are present in nature, as outlined in Chapter 2. The particular high-mannose oligosaccharide synthesised in this paper contains an 8-(methoxycarbonyl)octyl group at the anomeric centre rather than a chitobiose disaccharide, as would be found in naturally occurring high-mannose oligosaccharides (Figure 6.1).

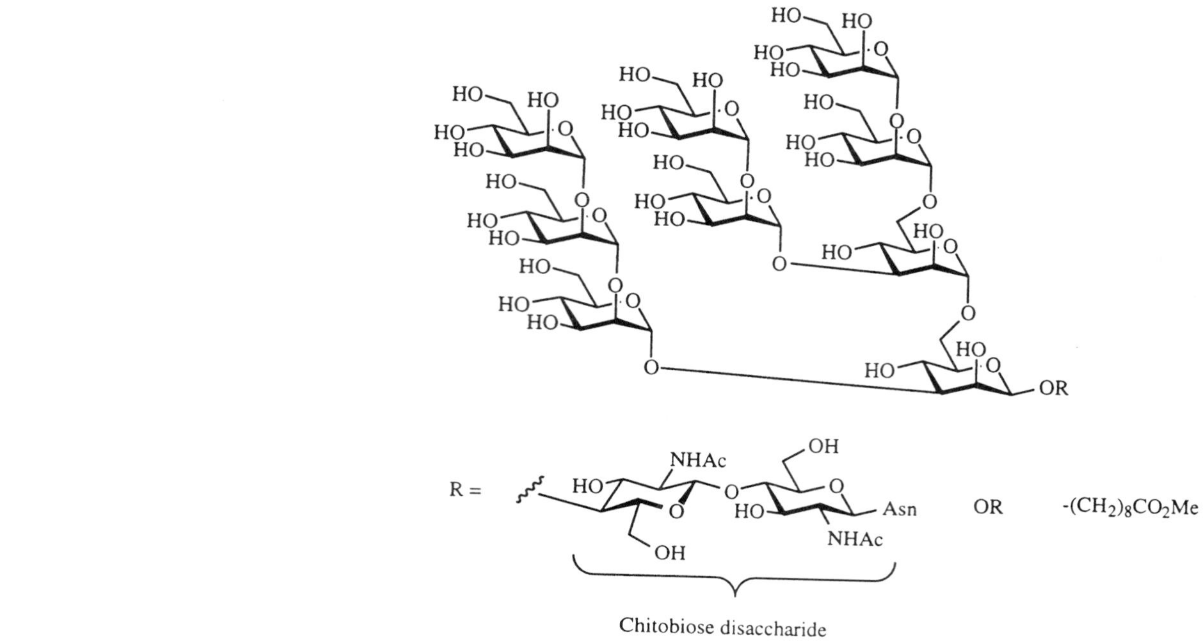

$R = $ Chitobiose disaccharide   OR $-(CH_2)_8CO_2Me$

Figure 6.1

The 8-(methoxycarbonyl)octyl group was incorporated at the anomeric centre so that the nonasaccharide could subsequently be attached to a protein or solid support for easy analysis of its biological properties.

The synthesis provides a working example of tuning the reactivity of glycosyl donors and acceptors by incorporating different protecting groups and anomeric groups within the carbohydrate building blocks. Retrosynthetic analysis of the nonasaccharide illustrates that a pentasaccharide and tetrasaccharide are required for its construction (Figure 6.2).

Furthermore, the branched pentasaccharide can itself be synthesised by reaction of an acceptor displaying two free hydroxyl groups with two equivalents of the disaccharide donor (Figure 6.3).

We will start by considering the synthesis of the tetrasaccharide and this is illustrated in Figure 6.4. Four different monomer units are employed in the synthesis, each with a different level of reactivity.

The difference in reactivity of the monosaccharide building blocks is a consequence of incorporating different anomeric leaving groups and different protecting groups within the donors and acceptors. Thus a fully benzylated glycosyl donor was initially reacted with a deactivated acceptor

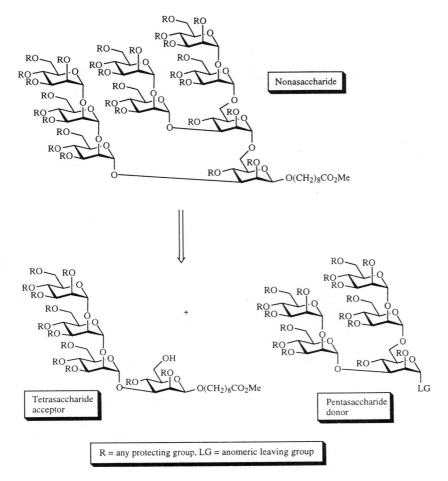

Figure 6.2

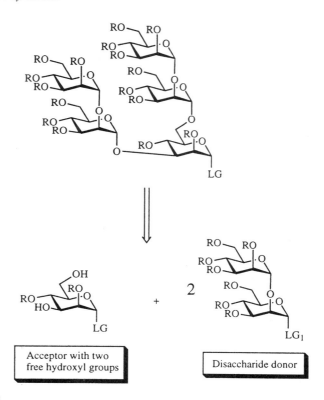

Figure 6.3

The reactivity of the building blocks was tuned by selecting specific protecting groups and anomeric groups.

(cyclohexane dimethyl acetal and benzoate protecting groups cause deactivation) to afford an α-linked disaccharide. This disaccharide was not isolated but reacted *in situ* with a less-reactive thioethyl acceptor to afford an α-linked trisaccharide in one pot.

It was found that this trisaccharide was so unreactive, due to the presence of the cyclohexane dimethyl acetal and benzoate protecting groups and the thioethyl group at the anomeric centre, that further activation with *N*-iodosuccinimide and triflic acid (NIS and TfOH) was impossible. Thus the thioethylpyranoside trisaccharide was converted into the more-reactive bromopyranoside trisaccharide donor prior to coupling with a further monosaccharide acceptor. This afforded the required protected tetrasaccharide in an overall yield of 40% from the monomer building blocks. Subsequent removal of the silyl protecting group with tetrabutylammonium fluoride then provided the tetrasaccharide acceptor ready for reaction with the pentasaccharide donor.

Synthesis of the pentasaccharide donor was also possible in an efficient manner using monosaccharide building blocks of modified reactivity.

Although both the donor and acceptor components required for synthesis of the disaccharide were phenylselenopyranosides, the acceptor was deactivated compared with the donor due to the cyclohexane dimethyl acetal protecting group. Once the disaccharide was assembled it was reacted *in situ* with a thioethyl acceptor displaying two free hydroxyl groups to furnish the required branched pentasaccharide. This was then coupled with the tetrasaccharide acceptor illustrated in Figure 6.4 under

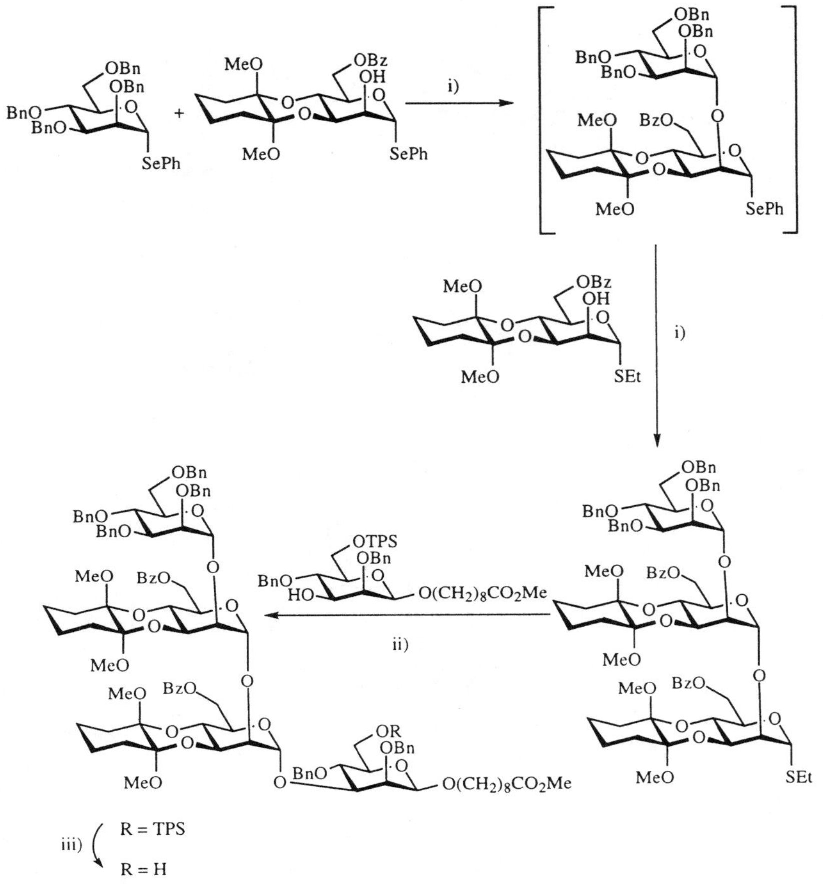

i) NIS, TfOH, 4Å molecular sieves; ii) 2,6-*tert*-butylpyridine, AgOTf, bromine, CH₂Cl₂; iii) TBAF, catalytic AcOH, THF

Figure 6.4

N-iodosuccinimide and triflic acid (NIS / TfOH) activation conditions to afford the fully protected nonasaccharide. In all cases, the glycosidation reactions selectively produced the α-linked glycosides which are thermodynamically favoured due to the anomeric effect. A series of high-yielding deprotection reactions were then performed to complete the synthesis of the required nonasaccharide (Figure 6.6).

The syntheses of some of the phenylselenopyranoside building blocks are highlighted below (Figure 6.7). These syntheses illustrate the utility of employing cyclohexane diacetal groups for the protection of *trans*-hydroxyl groups as well as the ability to selectively protect primary hydroxyl groups.

Stereoselective construction of the tetrasaccharyl cap portion of *Leishmania* lipophosphoglycan

The second example we have chosen to discuss illustrates the use of *n*-pentenyl glycosides and their derivatives for the synthesis of a tetrasaccharide which is a component of the cap region of the lipophosphoglycan

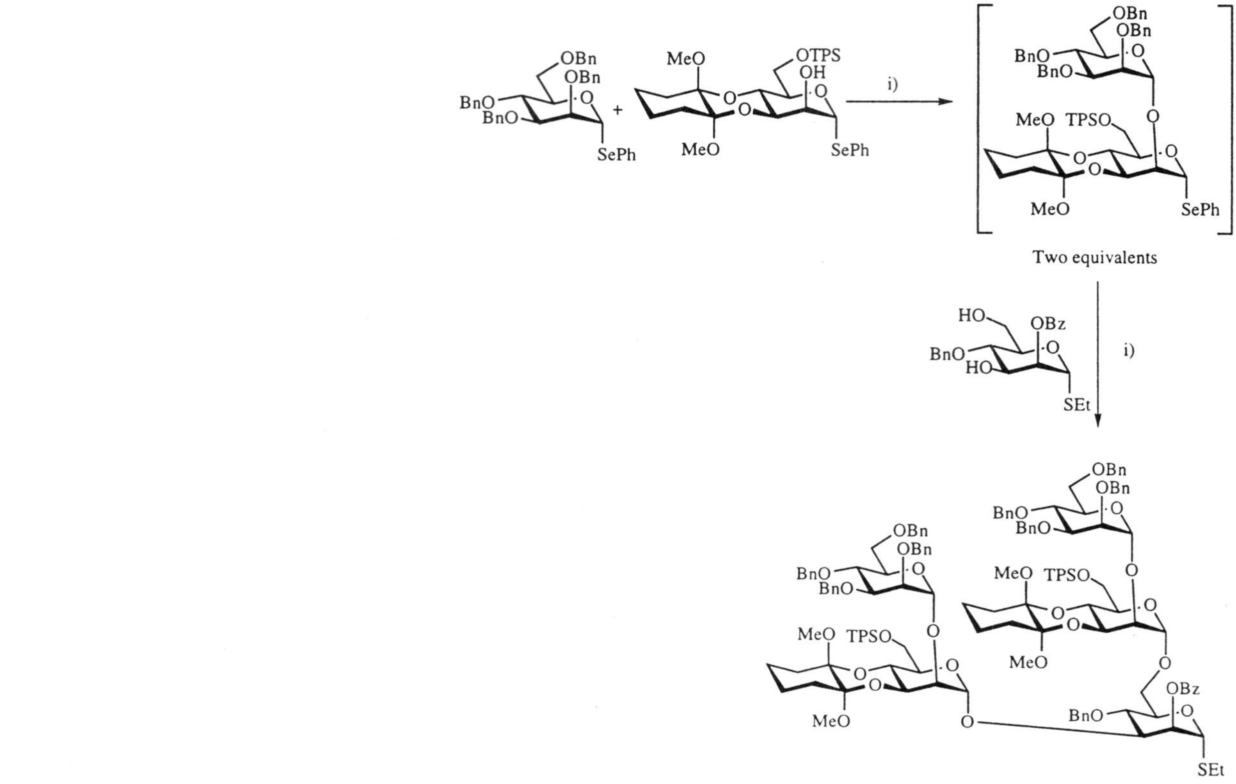

i) NIS, TfOH, 4Å ms, DCE, Ether

Figure 6.5

from the protozoan parasite *Leishmania*. These protozoan parasites are responsible for causing a tropical and subtropical disease called *Leishmaniasis*. There is evidence that the parasites' cell surface oligosaccharides play a key role in mediating the interactions which protect the parasite in the host's hydrolytic environment. Synthetic entry to these oligosaccharides, or fragments of these oligosaccharides, may allow for a better understanding of their biological roles. The oligosaccharides present on the *Leishmania* parasite can be divided into three classes, namely the cap, a repeating unit and the glycosylphosphatidylinositol (GPI) anchor oligosaccharides. This example will demonstrate the synthesis of the tetrasaccharyl cap oligosaccharide rather than of the repeating unit or the GPI anchor oligosaccharides.

The tetrasaccharide has been synthesised in both a linear and a convergent manner, but we will only discuss the more-efficient convergent synthesis in this Chapter. A possible retrosynthetic analysis of the targeted tetrasaccharide is highlighted in Figure 6.9, which illustrates that a suitably protected disaccharide donor and a suitably protected disaccharide acceptor are necessary for its synthesis.

We can envisage synthesising the disaccharide donor from a monosaccharide donor and a monosaccharide acceptor, both of which can be derived from the same *n*-pentenyl glycoside (Figure 6.10).

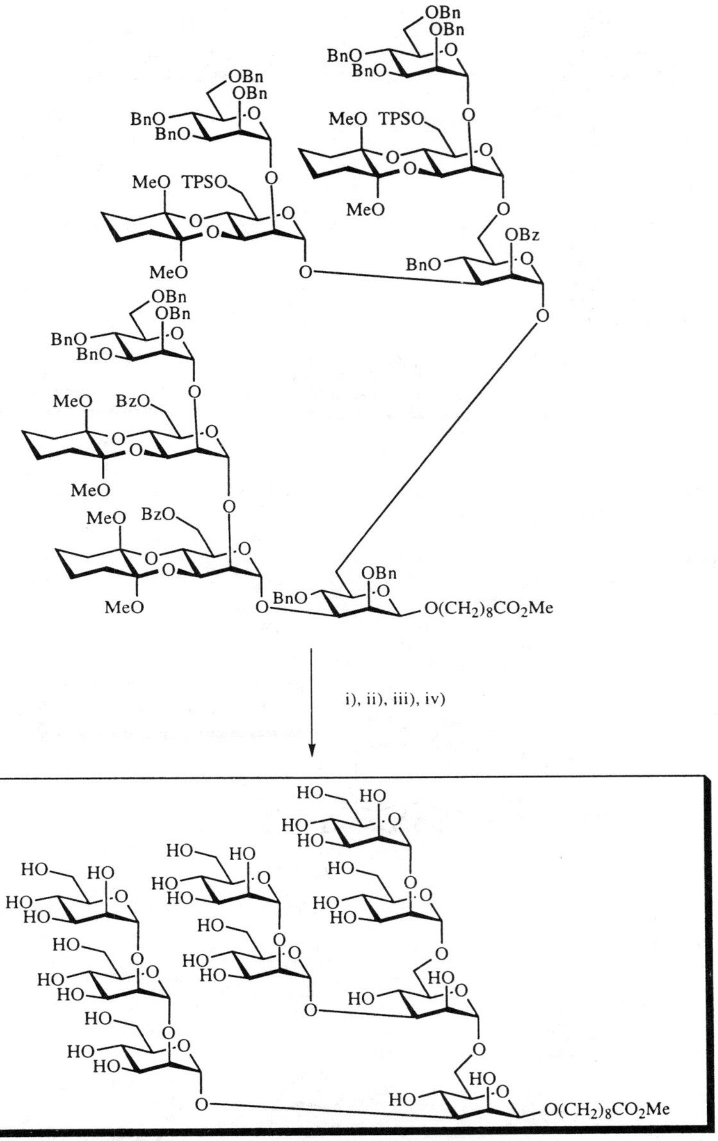

i) NaOMe, MeOH; ii) HF-pyridine; iii) TFA:H$_2$O (20 : 1); iv) Pd(OAc)$_2$, H$_2$

Figure 6.6

Likewise, the disaccharide acceptor requires prior synthesis of a mono-saccharide donor and an orthogonally protected monosaccharide acceptor (Figure 6.11).

This example of oligosaccharide synthesis is particularly interesting since it demonstrates a further utility of *n*-pentenyl glycosides which was not covered in Chapter 4. *n*-Pentenyl glycoside donors can be converted into glycoside acceptors by temporarily masking the *n*-pentenyl glycoside. This is easily achieved by treating the *n*-pentenyl glycoside with bromine, to allow formation of the dibromo analogue, which is no longer reactive under standard *n*-pentenyl-glycoside-activating conditions. If a mixture

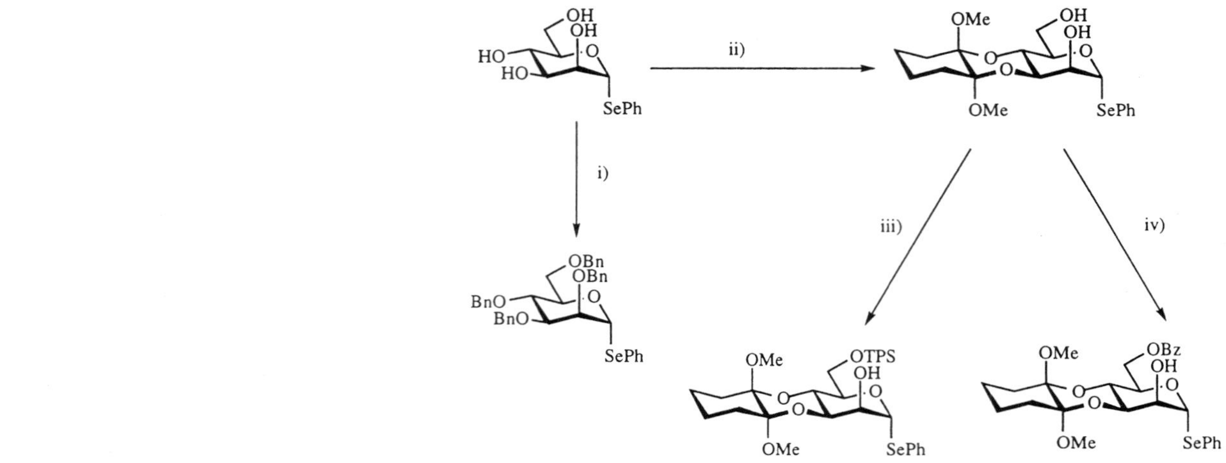

i) NaH, BnBr, DMF; ii) 1,1,2,2-tetramethoxycyclohexane, MeOH, CSA, (MeO)$_3$CH; iii) TPSCl, imidazole; iv) (Bu$_3$Sn)$_2$O, BzCl

Figure 6.7

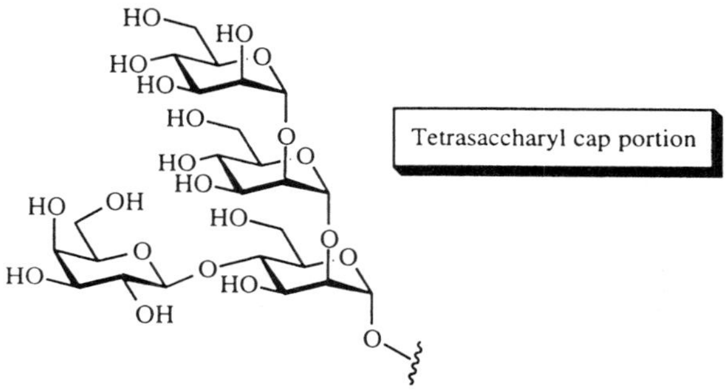

Tetrasaccharyl cap portion

Figure 6.8

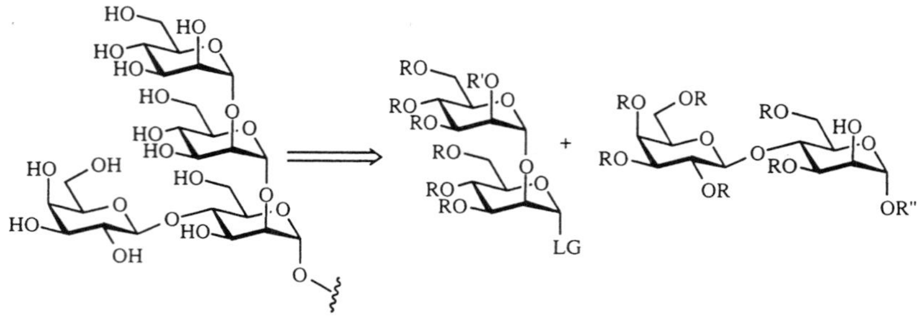

Figure 6.9

of an *n*-pentenyl glycoside donor and a dibromo acceptor is then treated under *n*-pentenyl-glycoside-activating conditions, only the disaccharide derived from heterocoupling results. After glycosidation, it is possible to re-form an *n*-pentenyl glycoside donor for further reactions via reductive debromination conditions. This is exemplified in Figure 6.12.

Thus a specific *n*-pentenyl glycoside can be tailored to act as an acceptor or as a donor by employing the bromination–debromination sequence of reactions.

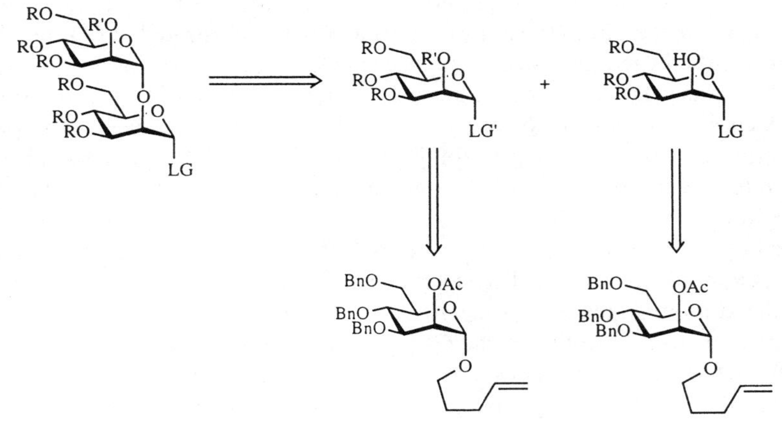

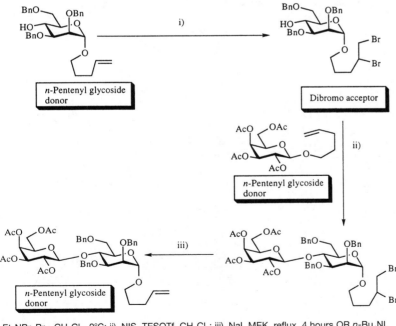

Figure 6.10

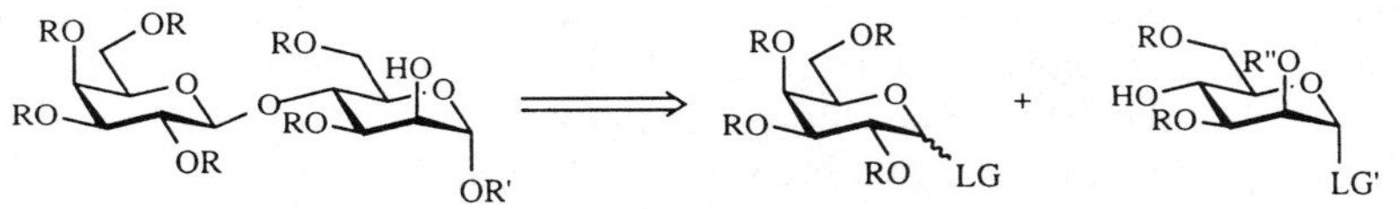

Figure 6.11

n-Pentenyl glycoside donors can be converted into glycoside acceptors by temporarily masking the reactivity of the n-pentenyl glycoside.

i) Et$_4$NBr, Br$_2$, CH$_2$Cl$_2$, 0°C; ii) NIS, TESOTf, CH$_2$Cl$_2$; iii) NaI, MEK, reflux, 4 hours OR n-Bu$_4$NI, Zn dust, EtOH/EtOAc, sonicate, 17 hours

Figure 6.12

We will commence by considering the synthesis of the disaccharide acceptor precursor to the tetrasaccharide which is illustrated in Figure 6.13. This initially required access to a partially protected glycoside acceptor which was prepared from mannose in a series of standard transformations which we have previously introduced in Chapter 3. The benzylidene protecting group was employed for protection of the C-4 and C-6 hydroxyl groups, whilst the C-3 hydroxyl group was protected via a tin-mediated regioselective protection reaction. The remaining C-2 hydroxyl group was orthogonally protected prior to regioselective opening of the benzylidene acetal to afford a free hydroxyl group at C-4. This latter reaction was considered more suitable than the alternative acidic cyanoborohydride conditions due to the lability of the C-2 chloroacetate functionality. The donor characteristics of this *n*-pentenyl glycoside were then removed by conversion of the *n*-pentenyl glycoside to its dibromo analogue under the conditions introduced in Figure 6.12. This meant that glycosidation with a further *n*-pentenyl glycoside was possible under standard activating conditions to afford the required disaccharide in good yield. In this case glycosidation afforded the β-linked disaccharide due to neighbouring group participation by the acetate ester at C-2 of the *n*-pentenyl glycoside donor. Removal of the orthogonal chloroacetyl group was then achieved to afford the acceptor required for synthesis of the tetrasaccharide.

NGP allows entry to the β-linked disaccharide.

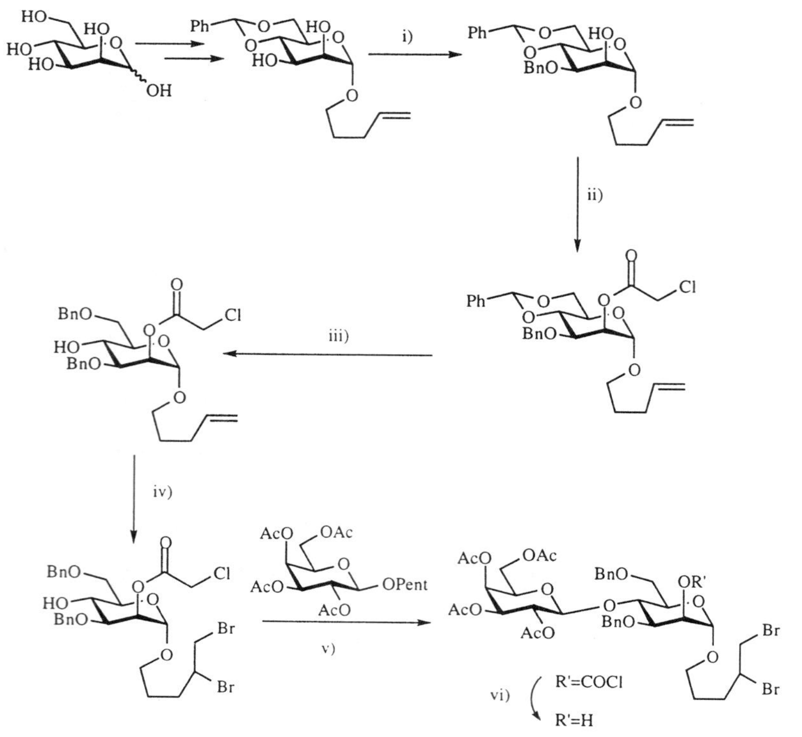

i) a) Bu$_2$SnO, MeOH, reflux; b) BnBr, DMF, 80–110°C; ii) (ClCH$_2$CO)$_2$O, pyr, 0–10°C; iii) Et$_3$SiH, CF$_3$CO$_2$H, CH$_2$Cl$_2$, 0–20°C; iv) Et$_4$NBr, Br$_2$, CH$_2$Cl$_2$, 0°C; v) NIS, TESOTf, CH$_2$Cl$_2$; vi) Thiourea, NaHCO$_3$, EtOH/EtOAc, reflux

Figure 6.13

The disaccharide donor required for synthesis of the tetrasaccharide was also prepared using an *n*-pentenyl glycoside donor and a dibromo acceptor. Conversion of the intermediate dibromo disaccharide to the analogous *n*-pentenyl glycoside disaccharide was again achieved under reductive debromination conditions (Figure 6.14).

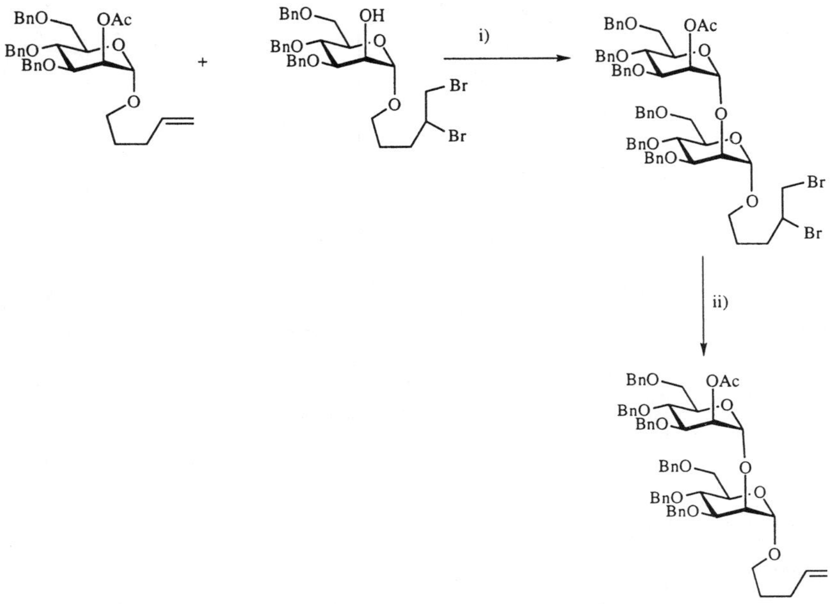

i) NIS, TESOTf, CH$_2$Cl$_2$; ii) *n*-Bu$_4$NI, Zn dust, EtOH/EtOAc, sonicate

Figure 6.14

Neighbouring group participation affects the stereoselectivity of the glycosidation step. As we saw in Chapter 5, mannopyranosides with an ester group at C-2 afford α-linked oligosaccharide. All that remained to complete the assembly of the tetrasaccharide was reaction of the dibromo acceptor with the *n*-pentenyl glycoside donor under typical *n*-pentenyl-glycoside-activating conditions (Figure 6.15).

This synthesis has therefore exemplified the utility of *n*-pentenyl glycosides for the synthesis of complex oligosaccharides. Furthermore, we have seen that *n*-pentenyl glycosides need not act solely as glycosyl donors, but are also useful precursors to dibromo glycosyl acceptors.

A highly convergent synthesis of the tetragalactose moiety of the glycosyl phosphatidyl inositol anchor of the variant surface glycoprotein of *Trypanosoma brucei*

Our third example considers a highly convergent synthesis of a partial structure of the glycosyl phosphatidyl inositol (GPI) anchor of the variant surface glycoprotein of *Trypanosoma brucei*. GPI anchors are involved in the covalent attachment of the C-termini of certain proteins to cell surfaces and have attracted much attention with regard to synthesis in recent years. Interestingly, this synthesis uses peracetylated galactose as

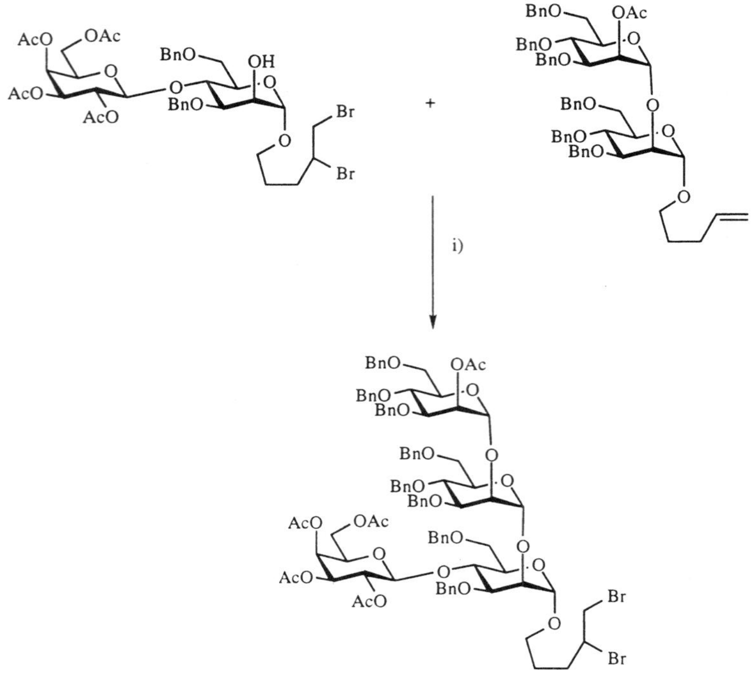

i) NIS, TESOTf, CH$_2$Cl$_2$

Figure 6.15

the precursor to both the sulfoxide donor and the thiophenyl acceptor components, which are necessary for full assembly of the tetrasaccharide (Figure 6.16).

The thiophenyl glycoside is accessed from commercially available per-acetylated galactose via reaction with thiophenol under Lewis acid activation conditions. The β-thiopyranoside results due to neighbouring group participation. Complete removal of the acetate groups is then achieved by

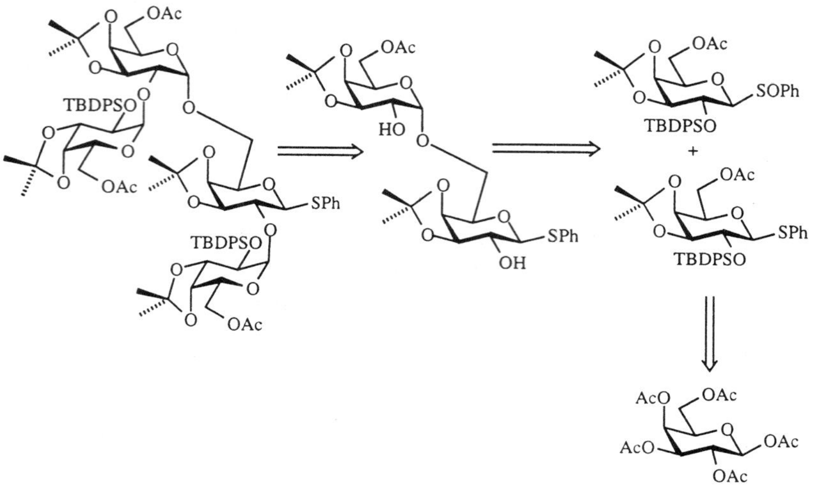

Figure 6.16

treatment with sodium methoxide to allow subsequent protection of the *cis*-hydroxyl groups as the acetonide. In order to produce an orthogonally protected acceptor, the more reactive primary hydroxyl group of this intermediate is protected as the acetate ester before protecting the remaining hydroxyl group at C-2 as its silyl ether. The acetate moiety at C-6 can then be removed to furnish the required thiophenylglycoside acceptor. Alternatively, the acetate moiety can be left intact and the thiophenylglycoside oxidised with *m*-CPBA to afford the corresponding sulfoxide donor (Figure 6.17).

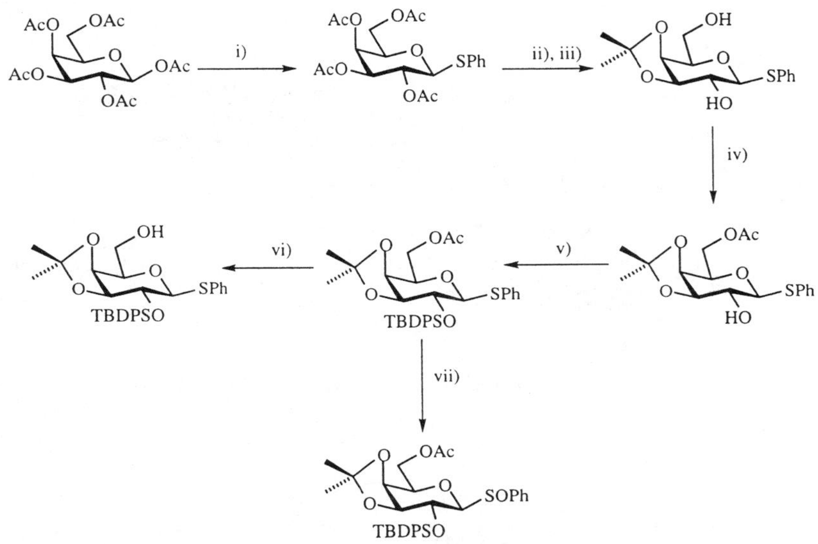

i) PhSH, BF$_3$.OEt$_2$, CH$_2$Cl$_2$, room temperature; ii) MeONa, MeOH, room temperature; iii) DMP, acetone, H$^+$, room temperature; iv) AcCl, collidine, CH$_2$Cl$_2$,–78°C; v) TBDPSCl, imidazole, DMAP, DMF; vi) MeONa, MeOH; vii) *m*-CPBA, CH$_2$Cl$_2$

Figure 6.17

Chemoselective glycosidation is effected using activating conditions which only affect the sulfoxide donor. Deprotection of the two silyl ethers in the resulting disaccharide is achieved by treatment with tetrabutylammonium fluoride to afford a disaccharide acceptor with a free hydroxyl group at each of the monosaccharide units. Reaction of this acceptor with a further two equivalents of the sulfoxide donor then allows entry to the required branched tetrasaccharide (Figure 6.18).

Synthesis of sialyl-Lewis^x

Many syntheses of sialyl-Lewis^x exist in the literature, but in the final example of this chapter we wish to discuss one strategy which demonstrates the utility of glycals for oligosaccharide assembly.

We have already seen in Chapter 4 that glycals are particularly useful donor molecules, since their activation and reaction with nucleophiles affords oligosaccharides with a free hydroxyl group at C-2. Under alternative activating conditions, it is possible to install a free amine group at

Orthogonal glycosidation strategies can be employed as the donor and acceptor possess different anomeric groups.

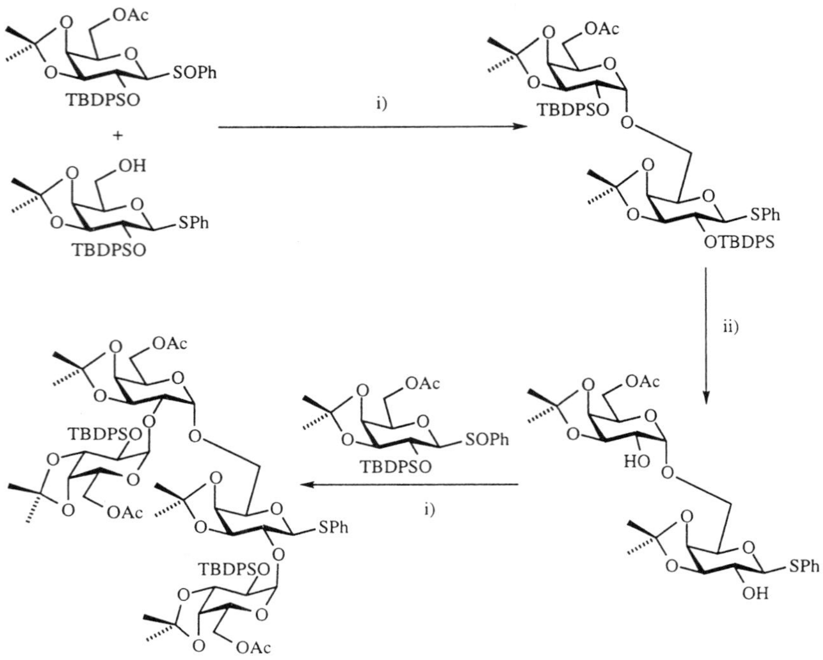

i) Tf$_2$O, 2,6-di-*tert*-butyl-4-methylpyridine, Et$_2$O:CH$_2$Cl$_2$, 3 : 1,–78°C; ii) TBAF, THF

Figure 6.18

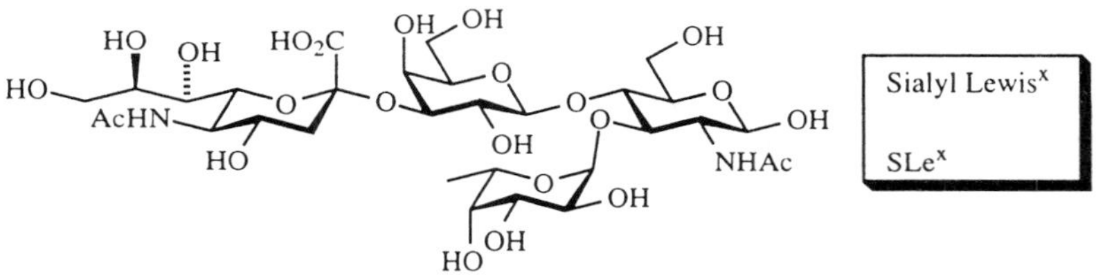

Figure 6.19

C-2, rather than a free hydroxyl group, and this is particularly useful for the synthesis of glucosamine-containing oligosaccharides. For example, when glycals are treated with a source of iodonium ions and nitrogen nucleophiles, iodosulfonamides are afforded which, upon treatment with stannyl alkoxides, can rearrange to afford glucosamine precursors (Figure 6.20).

Glycals are useful precursors of glucosamine derivatives.

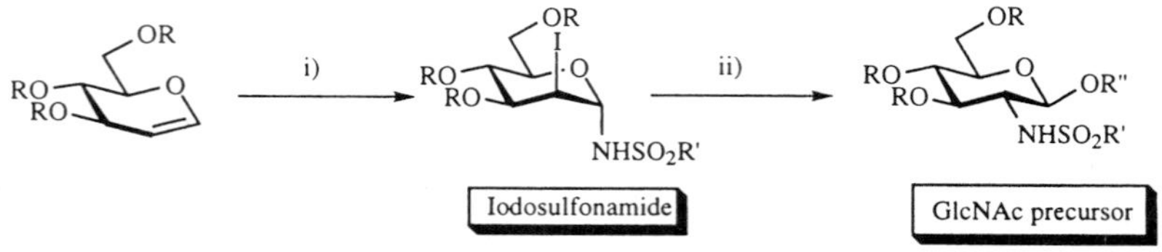

i) H$_2$NSO$_2$R′, I(*sym*-coll)$_2$ClO$_4$, CH$_2$Cl$_2$; ii) Bu$_3$SnOR″, AgOTf, THF, –78°C

Figure 6.20

In this particular synthesis of sialyl-Lewis[x], a glycal is incorporated within the growing oligosaccharide at an early stage of the synthesis, but the donor characteristics of the glycal are not called upon until a later stage. A retrosynthetic analysis of sialyl-Lewis[x] showing the key molecules necessary for its synthesis is highlighted below (Figure 6.21). A range of anomeric groups are incorporated within these key molecules and each class has already been identified as useful donors in Chapter 4. It is worth noting, however, that only donors which are activated under mild reaction conditions are employed in this synthesis due to the lability of the glycal functionality.

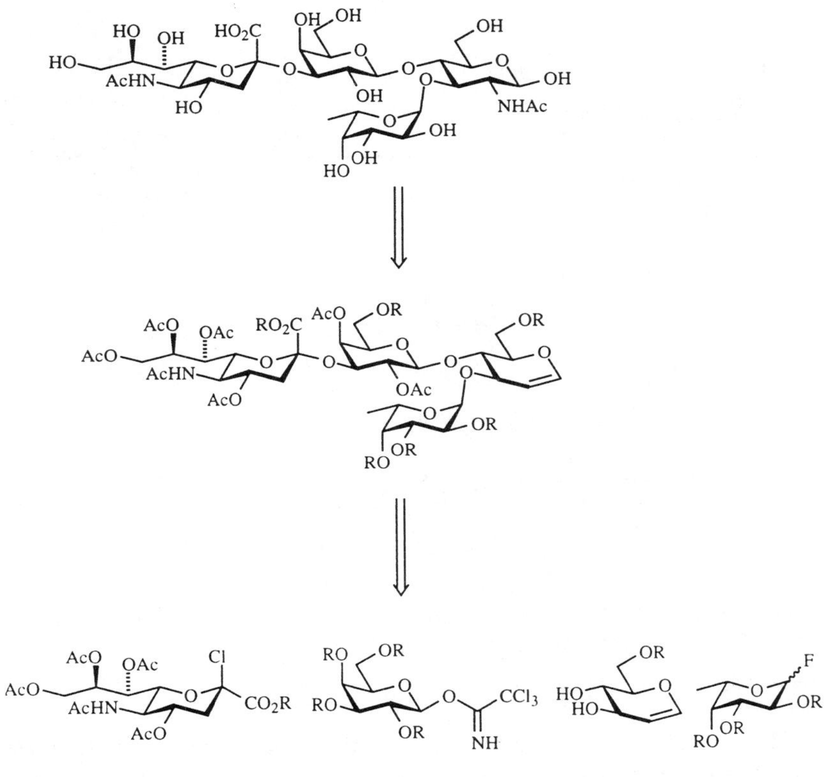

Figure 6.21

The synthesis commences with a regioselective glycosidation reaction of a glycal acceptor with a fluoropyranoside donor, with the donor being used as a mixture of anomers (Figure 6.22).

Fucosylation occurs predominantly at the less-hindered, and more-reactive, C-3 hydroxyl group of the glycal. The glycosyl fluoride was activated for glycosidation under Lewis acid reaction conditions, but a base, 2,6–di-*tert*-butylpyridine (DTBP), was added to the reaction medium to minimise decomposition of the acid labile glycal. Interestingly, only the required α-linked disaccharide was obtained in this manner. The free hydroxyl group of the furnished acceptor was then coupled with a trichloroacetimidate donor (mixture of anomers) to afford the core trisaccharide of Lewis[x]. Deprotection of the benzoyl esters on the galactose

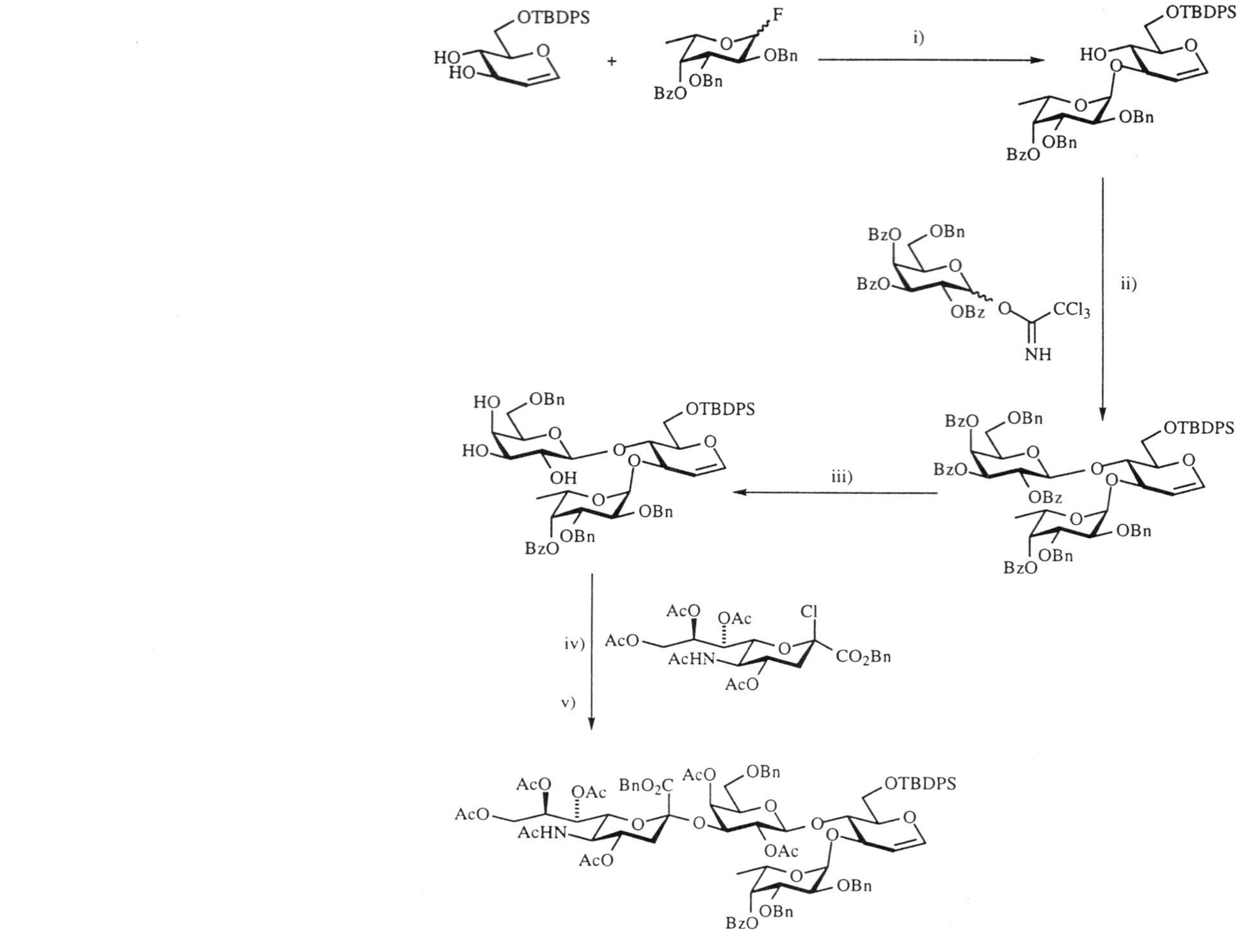

i) AgClO$_4$, SnCl$_2$, DTBP, Et$_2$O; ii) BF$_3$.OEt$_2$, CH$_2$Cl$_2$, −78°C; iii) NaOMe, MeOH; iv) AgOTf, DTBP, CaSO$_4$, THF, −78°C to −15°C; v) Ac$_2$O, pyr

Figure 6.22

residue was effected with sodium methoxide in methanol, but under these conditions the fucosyl benzoate ester remained intact. A further regio-selective coupling reaction of the trisaccharide triol with a chloropyranoside was then effected to afford the core structure of sialyl-Lewis^x. A proton scavenger was again required during the glycosidation reaction to minimise decomposition of the glycal, and the desired α-(2, 3)-linked sialoside was isolated in 40% yield from the precursor trisaccharide after protection of the resulting diol with acetic anhydride in pyridine.

This strategy was also employed for the synthesis of an alternative sialyl-Lewis^x precursor which differed from that described above only in the protection of the fucosyl portion of the tetrasaccharide. This tetrasaccharide was elaborated to afford sialyl-Lewis^x by employing the iodosulfonamide protocol introduced above. Thus the glycal was treated with 2-(triethylsilyl)ethanesulfonamide in the presence of iodonium bis(*sym*-collidine) perchlorate to afford the sulfonamide suitable for

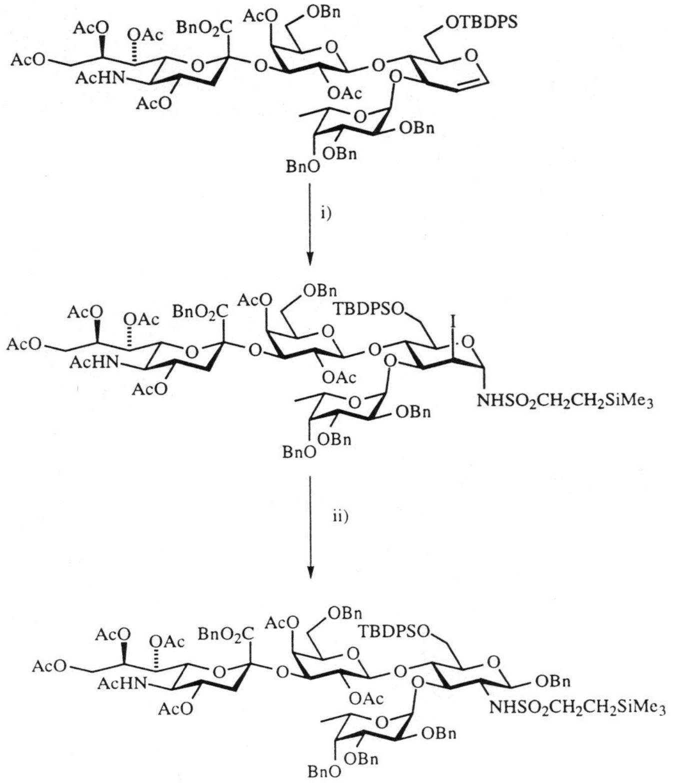

i) $H_2NSO_2CH_2CH_2SiMe_3$, $I(sym\text{-coll})_2ClO_4$,$CH_2Cl_2$; ii) Bu_3SnOBn, AgOTf, THF, $-78°C$

Figure 6.23

rearrangement to sialyl-Lewis[x]. The rearrangement reaction was effected by treatment of the sulfonamide with a stannyl alkoxide and this afforded the β-linked benzyl glycoside in 64% yield. A series of deprotection steps were then necessary to yield Sialyl-Lewis[x] (Figure 6.23).

Further use of the iodosulfonamide methodology

We will finish this Chapter by illustrating a further use of the iodo-sulfonamide methodology introduced above. It has been shown that stannyl glycoxides are also able to effect the rearrangement reactions of the iodosulfonamides. In these cases, even more complex oligosaccharides can be prepared in high yield from the glycal precursors. For example, Figure 6.24 illustrates elaboration of a sulfonamide derived from a tetrasaccharide glycal to a penta- and hexasaccharide.

Since these oligosaccharides also incorporate glycals there is scope for developement of an iterative iodosulfonamide-stannyl glycoside glycosidation protocol.

Further reading

1. Grice, P., Ley, S.V., Pietruszka, J., Osborn, H.M.I., Priepke, H.W.M., Warriner, S.L. (1997). A new strategy for oligosaccharide assembly

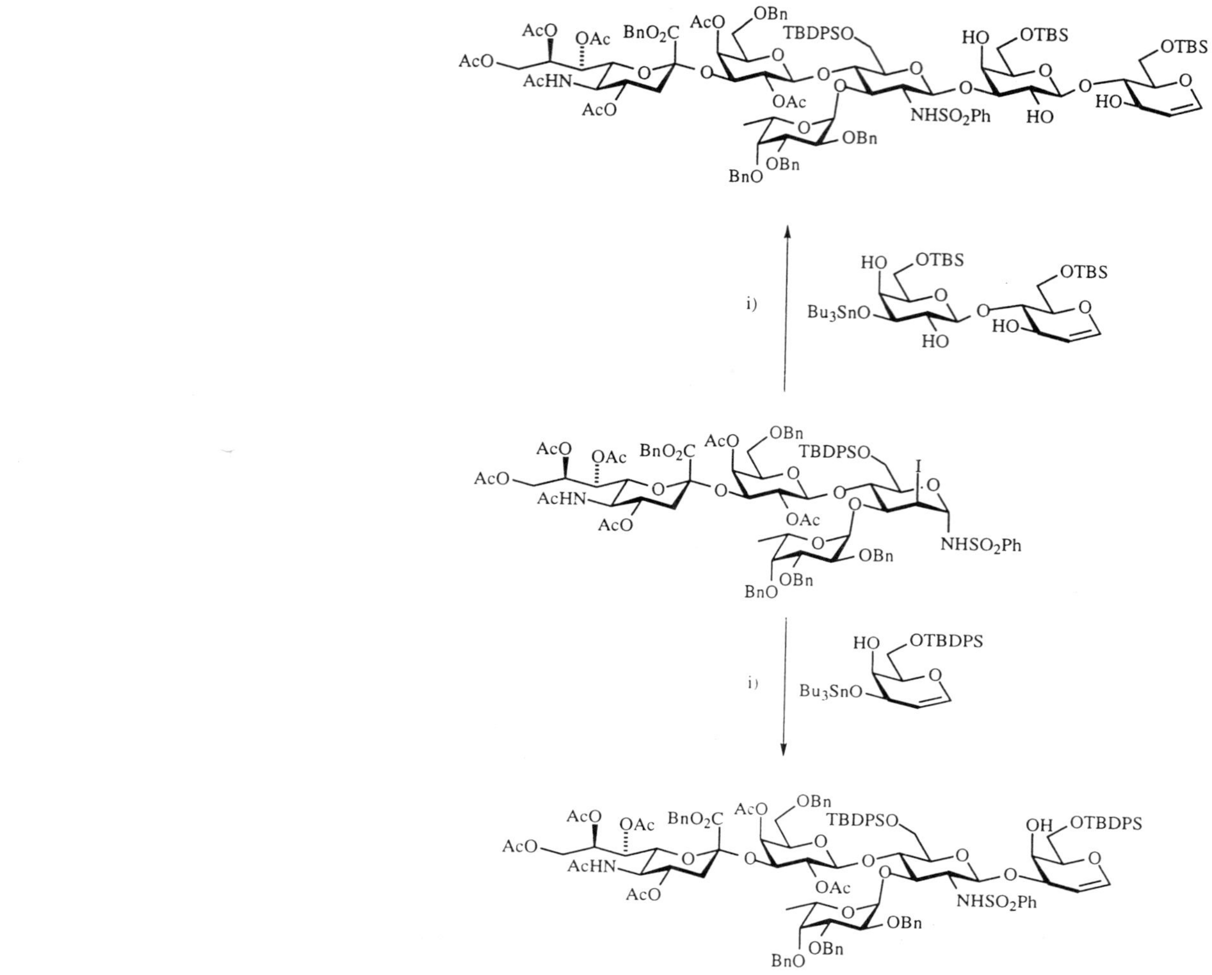

i) AgBF$_4$, THF, −78°C to room temperature, 3 days

Figure 6.24

exploiting cyclohexane-1,2-diacetal methodology : an efficient synthesis of a high mannose type nonasaccharide. *Angewandte Chemie International Edition in English*, **3**, 431–440.

2. Arasappan, A., Fraser-Reid, B. (1996). *n*-Pentenyl glycoside methodology in the stereoselective construction of the tetrasaccharyl cap portion of *Leishmania* lipophosphoglycan. *Journal of Organic Chemistry*, **61**, 2401–2406.

3. Khiar, N., Martin-Lomas, M. (1995). A highly convergent synthesis of the tetragalactose moiety of the glycosyl phosphatidyl inositol anchor of the variant surface glycoprotein of *Trypanosoma brucei*. *Journal of Organic Chemistry*, **60**, 7017–7021.

4. Danishefsky, S.J., Gervay, J., Peterson, J.M., McDonald, F.E., Koseki, K., Oriyama, T., Griffith, D.A., Wong, C.-H., Dumas, D.P. (1992).

Remarkable regioselectivity in the chemical glycosylation of glycal acceptors : a concise solution to the synthesis of Sialyl-Lewis[x] glycal. *Journal of the American Chemical Society*, **114**, 8239–8331.

5. Danishefsky, S.J., Koseki, K., Griffith, D.A., Gervay, J. Peterson, J.M., McDonald, F.E., Oriyama, T. (1992). Azaglycosylation of complex stannyl alkoxides with glycal-derived iodo sulfonamides: a straight-forward synthesis of Sialyl-Lewis[x] antigen and other oligosaccharides domains. *Journal of the American Chemical Society*, **114**, 8331–8333.

6. Danishefsky, S.J., Gervay, J., Peterson, J.M., McDonald, F.E., Koseki, K., Griffith, D.A., Oriyama, T., Marsden, S.P. (1995). Application of glycals to the synthesis of oligosaccharides—convergent total synthesis of the Lewis[x] trisaccharide and Sialyl Lewis[x] antigenic determinant and higher congeners. *Journal of the American Chemical Society*, **117**, 1940–1953.

7 Alternative strategies for synthesising oligosaccharides

So far we have concentrated on the chemical synthesis of carbohydrate acceptors, donors and oligosaccharides. However, chemical methods are not the only methods available for synthesising oligosaccharides. Some alternative methods which also allow efficient synthesis of oligosaccharides are discussed in this chapter.

Enzymatic synthesis of oligosaccharides

The wide range of oligosaccharides which occur in nature are prepared via a series of biosynthetic reactions which use enzymes to perform specific reactions on a limited set of monosaccharides. The most common monosaccharides employed in these enzymatic pathways are glucose, galactose, N-acetylglucosamine, xylose, glucuronic acid, fucose, mannose and sialic acid.

Enzymes can be used to assemble oligosaccharides using a range of naturally occurring monosaccharide building blocks.

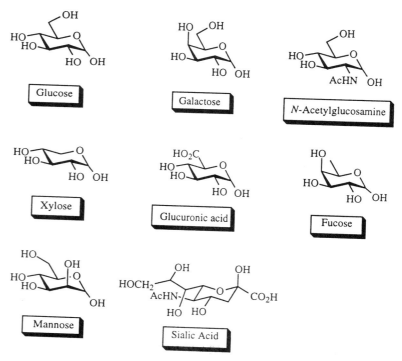

Figure 7.1

Much attention has focused upon the isolation and purification of these enzymes so that they can be used for the synthesis of carbohydrates in the laboratory. There are a number of advantages associated with enzyme-mediated reactions, including:

(1) The high levels of regioselectivity obtained in the reactions. This means that entry to only the regioisomer of choice is routinely effected, which bypasses the need for elaborate protecting-group strategies. This is therefore efficient both in terms of the number of synthetic steps required to effect the preparation of the target compound, and in terms of purification procedures.
(2) The high levels of stereoselectivity imparted. We can choose whether to access the α- or β-anomer of a specific target by carefully selecting which enzyme is employed in a strategy. Only rarely are mixtures of anomers obtained, which again simplifies purification procedures.

However, the use of enzymes is not without disadvantages and it is always necessary to balance the merits of a chemical strategy with those of an enzymatic strategy for each individual synthetic process. Some problems which are often encountered in enzymatic processes include:

(1) The rarity and hence relative expense of certain enzymes. Although biotechnologists have developed methods for over-expressing certain enzymes, improving their availability, only certain enzymes are commercially available at economical rates.
(2) Solubility problems. For certain substrates it is beneficial to perform an enzymatic reaction in organic solvents, and yet enzymes are often deactivated in organic solvents.
(3) Scale-up problems. In order to obtain useful quantities of oligosaccharides, large quantities of enzymes are often required. The availability and price of the enzymes required is often a limiting factor in such strategies. This problem has been overcome to a certain extent by attaching the enzyme to a polymer and using the polymer-bound enzyme as an aqueous suspension. In this way, enzyme deactivation throughout the reaction is diminished and the enzyme can easily be filtered and reused at the end of the synthesis. One example of this technique which has been reported by Whitesides involves the immobilization of enzymes by the condensation co-polymerisation of a water-soluble, functionalized prepolymer (PAN, poly(acrylamide)-*co*-*N*-acryloxysuccinimide), a low-molecular-weight α,ω-diamine (TET, triethylenetetramine) and the enzyme in neutral buffered aqueous solution. The diamine is reacted with the active ester groups of PAN to cross-link the polymer chain and form an insoluble gel connected through amide groups. The reactive amine functional groups of the enzyme are then reacted with the active esters on the poly(acrylamide) gel to covalently link the enzyme and polymer via amide bonds (Figure 7.2).

As more and more enzymes are being identified and cloned, the attractiveness of employing these enzymes in synthetic strategies is increasing.

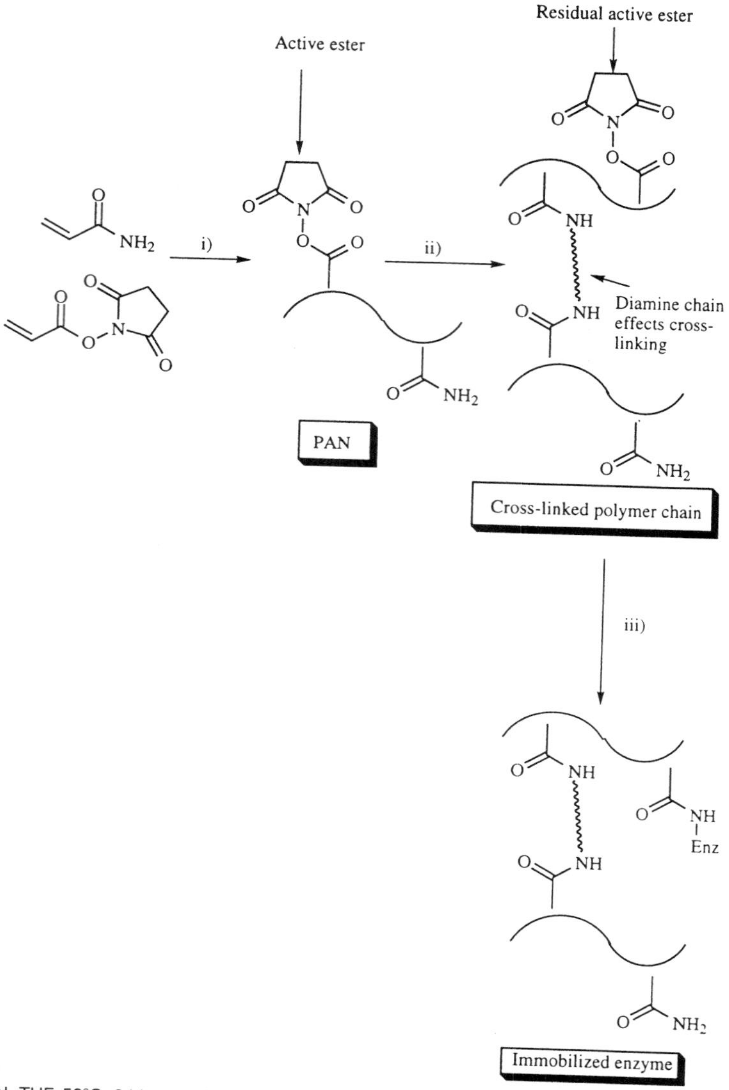

i) AIBN, THF, 50°C, 24 hours; ii) $H_2NCH_2CH_2NHCH_2CH_2NHCH_2CH_2NH_2$ (TET); iii) Enz–NH$_2$, active-site protective agents, pH 7.5, 60 minutes

Figure 7.2

Two main approaches have been studied for the enzymatic preparation of oligosaccharides, namely the use of glycosyl transferases and glycosyl hydrolases.

Glycosyl transferases

In cells, the formation of oligosaccharides occurs via the Leloir pathway. A nucleotide–sugar complex acts as donor and a transferase enzyme is employed to catalyse the transfer of the sugar to an acceptor molecule, allowing access to oligosaccharides and the free nucleotide. The free nucleotide is then used to regenerate the nucleotide–sugar complex for further reactions. Many glycosyl transferase enzymes which are used in the Leloir pathway have been isolated and used in the laboratory for the synthesis of complex oligosaccharides. Glucose, galactose, *N*-acetylgalactosamine, xylose and glucuronic acid are usually transferred from uridine diphosphate donors, whilst fucose and mannose are transferred from guanosine diphosphate donors and sialic acid is transferred from a cytidine monophosphate donor (Figure 7.3).

One of the most commonly employed enzymes used in oligosaccharide assembly strategies is β-(1,4)-galactosyltransferase (EC 2.4.1.22/38), which

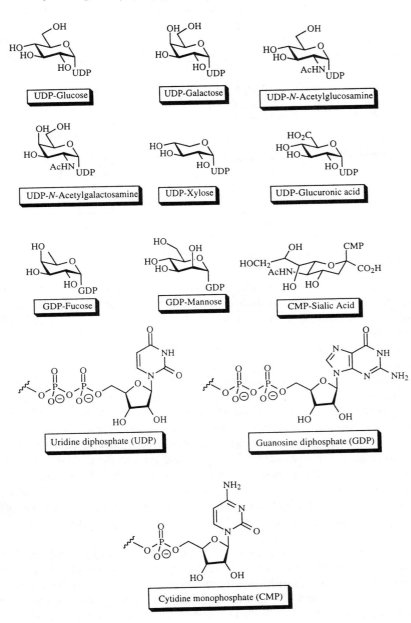

In nature, nucleotide–sugar complexes act as donors – transferase enzymes catalyse the transfer of the sugar to an acceptor molecule.

Figure 7.3

catalyses the transfer of a galactose molecule from UDP-gal to the C-4 hydroxyl group of GlcNAc. The *N*-acetyllactosamine structures thus produced are commonly found in the backbone regions of glycoproteins and glycolipids and are therefore interesting targets of biological importance. Each transferase enzyme is classified according to the nucleotide donor sugar employed in the synthetic strategy, the specific linkage formed in the new oligosaccharide, and the specific hydroxyl group of the acceptor to which the donor becomes attached. The enzyme effecting the transfer reaction is highly specific with respect to the donor and acceptor molecules, the position at which the new glycosidic bond is formed and the stereochemical outcome of the coupling reaction. However, certain modifications in the acceptor moiety are sometimes tolerated, which allows enzymatic preparation of unnatural oligosaccharides of biological interest.

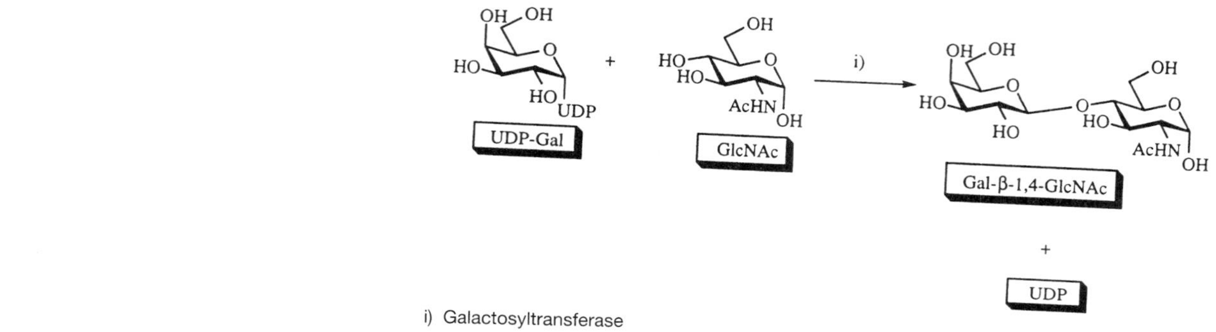

i) Galactosyltransferase

Figure 7.4

Alternative oligosaccharide targets of current importance are those displaying sialic acid or fucose residues at their termini, since these oligosaccharides have been demonstrated to play key roles in cell–cell recognition processes. A range of sialyl- and fucosyltransferase enzymes have been isolated which are responsible for the addition of sialic acid and fucose residues within biosynthetic pathways, namely α-(2,3)- and α-(2,6)-sialyltransferases, and α-(1,3)-fucosyltransferase.

Both the nucleotide donor molecules and the glycosyltransferases required for the above reactions are commercially available. Although the nucleotide donor molecules are laborious to synthesise chemically, they are easily prepared via enzymatic methods. Recent strategies have taken advantage of this discovery and employ enzyme recycling systems for the regeneration of the nucleotide–donor molecules in situ. In this way, only catalytic amounts of the nucleotide donor are required for the synthesis, and the nucleotide which is produced is constantly used for regeneration of the nucleotide–sugar donor. An example of this strategy is shown below in Figure 7.5.

In this example, galactosyl transferase (GT) initially catalyses the transfer of galactose to the C-4 hydroxyl group of *N*-acetylglucosamine to generate a disaccharide and an equimolar quantity of uridine diphosphate (UDP). The UDP nucleotide is enzymatically phosphorylated to form uridine triphosphate (UTP) using enolpyruvate phosphate in the presence of a further enzyme, pyruvate kinase (PK). A further transferase

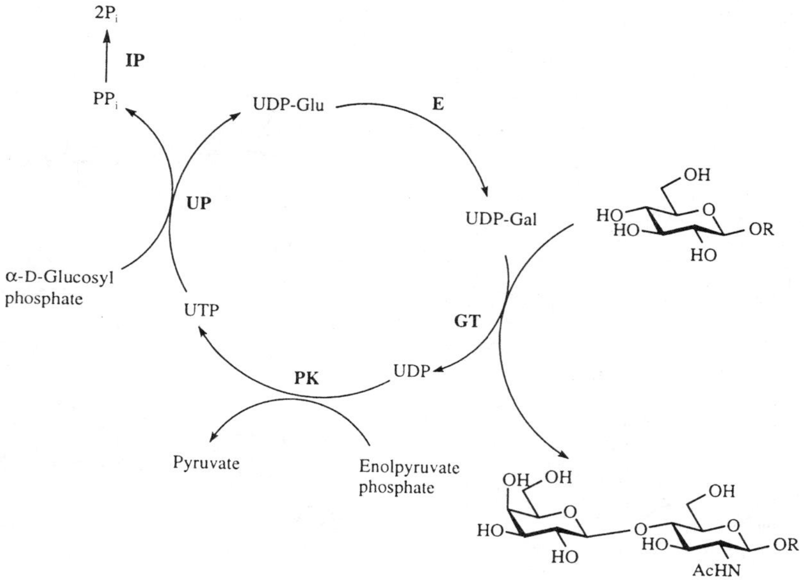

Enzyme recycling systems can be used to regenerate the nucleotide–donor molecules *in situ*.

Figure 7.5

enzyme, UDP-pyrophosphorylase (UP), then catalyses the synthesis of the nucleotide–sugar donor molecule, UDP–Glu, from the previously formed UTP and α-D-glucosyl phosphate. This latter reaction is reversible, hence the pyrophosphate by-product (PP$_i$) must be removed from the reaction medium to inhibit the reverse reaction. This is achieved by hydrolysing the PP$_i$ to inorganic phosphate with a further enzyme, inorganic pyrophosphatase (IP). The final reaction required to regenerate the nucleotide–galactose donor molecule is an epimerisation reaction which is again achieved enzymatically using an epimerase enzyme. All of the enzymes required for this process are commercially available.

Sialic acid residues are introduced in similar pathways using cytidine monophosphate-*N*-acetylneuraminic acid as the donor molecule in conjunction with suitable substrates and sialyl transferases (Figure 7.6).

These process are becoming more useful and more extensively applied as further transferase enzymes are identified and cloned. For example, it has recently become possible to address a problem introduced in Chapter 5, namely the synthesis of *N*-linked core oligosaccharides containing the β-mannosyl linkage. This has proved possible in excellent yield using an immobilised recombinant β-mannosyltransferase enzyme, which allows transfer of a mannose unit from a GDP–Man nucleotide donor to a chitobiose acceptor (Figure 7.7).

Glycosyl hydrolases

A further class of enzymes used for oligosaccharide synthesis is the glycosyl hydrolases. Glycosyl hydrolases are particularly attractive for synthetic strategies, as they are relatively common and inexpensive. These enzymes catalyse the hydrolysis of glycosidic bonds in a two-step procedure as illustrated in Figure 7.8.

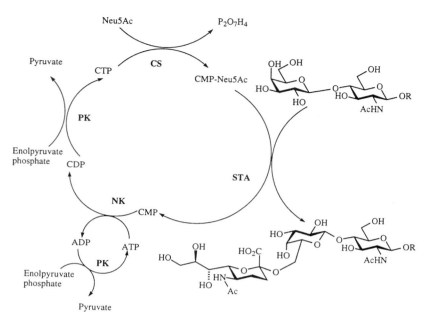

Figure 7.6

Figure 7.7

DONOR—LG + ENZYME—H ⟶ DONOR—ENZYME + LG-H

Glycosyl hydrolases catalyse the hydrolysis of glycosidic bonds.

DONOR—ENZYME + HO—H ⟶ DONOR—OH + ENZYME-H

Figure 7.8

In the first step of the reaction process, a glycosyl–enzyme complex is formed. The glycosyl group is then transferred from this complex to the hydroxyl group of water, releasing the hydrolysed glycosyl moiety and regenerating the enzyme catalyst. If hydroxylated derivatives other than water are employed as acceptors, for example alcohols or carbohydrates, then formation of new glycosidic bonds can be achieved.

As two configurational inversions occur at the anomeric centre, the anomeric configuration of the newly formed glycosidic bond is the same as that of the donor. Again, the reactions are regioselective but yields of products are often low and rarely exceed 30%. As the acceptors required are generally inexpensive, they can be used in excess to alleviate this problem (Figure 7.9).

DONOR—ENZYME + ACCEPTOR—H ⟶ DONOR–ACCEPTOR + ENZYME-H

The anomeric configuration of the newly formed glycosidic bond is the same as that of the donor.

Figure 7.9

It is not unusual to employ both hydrolase and transferase enzymes in a particular strategy, and whichever strategy proves most efficient should be employed. Indeed it is not unusual to use a combination of chemical and enzymatic methods for the synthesis of complex oligosaccharides. In these cases, linkages which are difficult to introduce chemically are introduced enzymatically.

Polymer-supported synthesis of oligosaccharides

We have seen that preparation of biologically important oligosaccharides typically requires multistep transformations involving repetitive protection–glycosidation–deprotection reactions. Chromatographic purification of intermediates is usually necessary at each stage of the synthesis. If such reactions could be performed on a solid support, the synthesis of complex oligosaccharides may be improved. This is because polymer-supported syntheses offer many advantages compared with the analogous solution phase reactions, including:

(1) Increased yields of reaction, due to the ability to add large excesses of each reagent to drive the reactions to completion.
(2) Increased speed of synthesis, as no purification processes are required.

The utility of solid supports has been recognised and developed for the synthesis of other classes of biologically important molecules, such as

The synthesis of complex oligosaccharides may be improved if the procedure is performed on a polymer support.

peptides and oligonucleotides. However, challenging problems are anticipated when polymer-supported approaches are applied to the syntheses of oligosaccharides. As we have seen, the preparation of a specific carbohydrate requires the stereospecific formation of each new glycosidic bond. Therefore solid-phase carbohydrate synthesis is currently less-well-developed than solid-phase peptide or oligonucleotide syntheses. However, many leading researchers are now concentrating on developing and improving methods for building oligosaccharides on solid supports.

Two main strategies can be envisaged for solid-phase oligosaccharide synthesis entailing either attachment of the acceptor or of the donor to the solid support. In the first strategy the acceptor is bound to the solid support, via any of its hydroxyl groups and a solution-based donor and promoter are added to allow coupling to occur. In the second approach, a glycosyl donor is bound to the solid support via any one of its hydroxyl groups. It is then reacted with a solution phase acceptor (Figure 7.10).

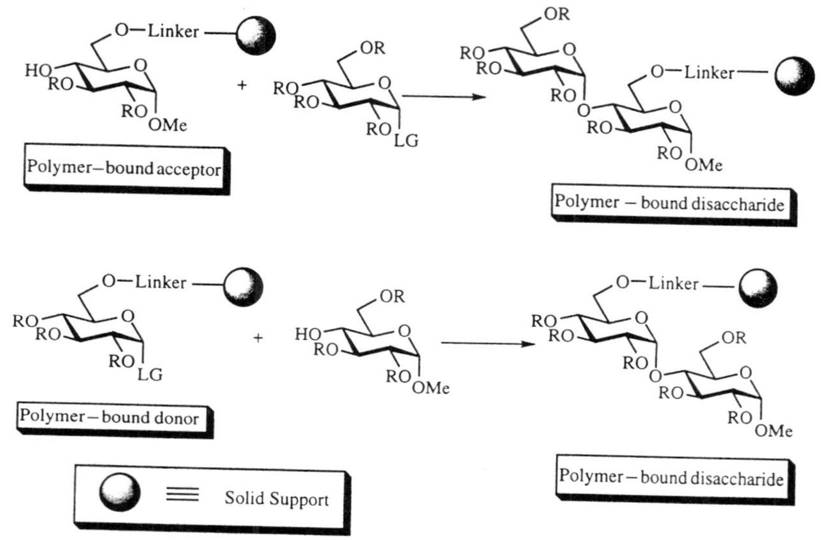

Figure 7.10

In either approach, elaboration of the protecting groups on the polymer bound oligosaccharide is necessary if further elaboration and elongation of the oligosaccharide is required. Both approaches have been developed and reported in the modern literature.

A range of glycopeptides have also been synthesised using solid supports. However, since polymer-supported synthesis of peptides is better established than that of carbohydrates, the majority of polymer-supported glycopeptide syntheses involve linking the glycopeptide precursor to the polymer support via the peptide moiety rather than via the carbohydrate moiety (Figure 7.11).

Choice of polymer

For any polymer-supported synthesis, a wide range of polymer supports are available and the most suitable polymer must be selected for the specific

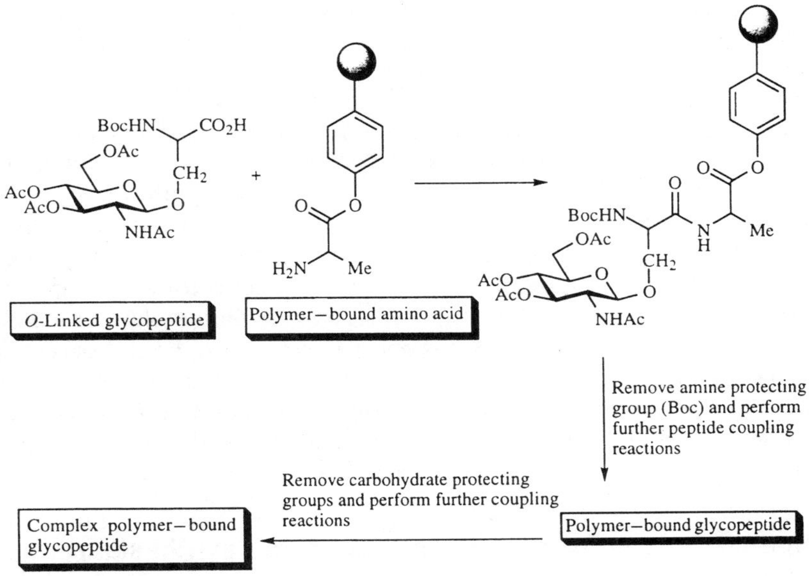

A range of glycopeptides have also been synthesised using solid supports.

Figure 7.11

reaction of interest. Furthermore, a method must be developed for linking the polymer to the carbohydrate which will be undergoing reaction. The Merrifield resin has been used extensively in the solid phase synthesis of oligosaccharides which, no doubt, is a consequence of its success in the solid-phase synthesis of peptides. A range of soluble polymer supports, which allow homogenous rather than heterogeneous reactions to occur, have also been employed. Analysis of the polymer-bound reaction products is then easy without the need for cleaving reaction intermediates and products from the solid support. Some popular soluble polymer supports which have been studied to date are highlighted in Figure 7.12.

Soluble polymer supports allow homogenous rather than heterogeneous reactions to occur.

Choice of linker system

A range of linking systems have also been developed which are compatible with reaction conditions necessary for oligosaccharide synthesis and allow efficient release of the newly formed oligosaccharide from the polymer at the end of the synthesis without any decomposition of the newly formed glycosidic bonds. A range of linkers has been developed which are removable under either acidic, basic, photolytic or enzyme-mediated conditions, which therefore produces a range of linkers compatible with

The saccharide unit is attached to the polymer via a linking system.

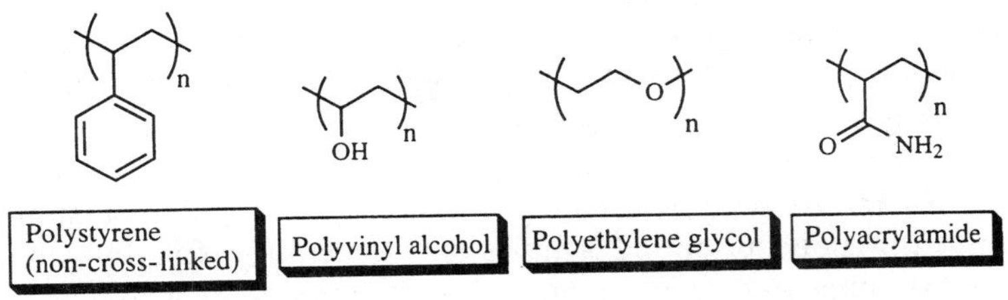

Polystyrene (non-cross-linked)   Polyvinyl alcohol   Polyethylene glycol   Polyacrylamide

Figure 7.12

all the reaction conditions that may be met in polymer-supported syntheses.

Synthesis of the oligosaccharide on the solid support

After the best polymer and linker systems have been selected for a particular strategy, attention must turn to assembly of the oligosaccharide itself. Many strategies reported for the *polymer-supported* syntheses of oligo-saccharides elaborate upon recent successful developments in the *solution* phase syntheses of oligosaccharides. Hence the more widely used anomeric groups, glycosidation conditions and tactics for controlling the geometry of newly formed bonds developed for solution-phase reactions have been applied to polymer-supported strategies. Trichloroacetimi-dates, *n*-pentenyl glycosides, anomeric sulfoxides and glycals are some of the most widely utilised donors for polymer-supported oligosaccharide strategies. Outlined below are just a few examples of polymer-supported syntheses which employ these intermediates.

Specific examples of polymer-supported synthesis of oligosaccharides

Synthesis of pentasaccharides containing repeated Glu-α-(1,6)-linkages and Man-α-(1,2)-linkages

Polymer-bound alkyl thiol acceptors have been used in conjunction with solution-phase trichloroacetimidate donors for the syntheses of penta-saccharides containing repeated Glu-α-(1,6)-linkages or Man-α-(1,2)-linkages via the iterative process outlined in Figure 7.13. In this example, the carbohydrate is linked to the polymer via reaction of the reactive alcohol or thiol moiety of the polymer-bound linker system with a mono-saccharide trichloroactimidate. In this way the monosaccharide becomes bound to the polymer via its anomeric position.

The polymer-bound monosaccharide produced has orthognal protect-ing groups which are central to this particular synthesis. In other words, an acetate group is employed as a temporary orthogonal protecting group for the C-2 hydroxyl group. This allows deprotection of only this C-2 hydroxyl group upon treatment of the polymer-bound monosaccharide with sodium methoxide. The polymer-bound monosaccharide is therefore converted to a polymer-bound acceptor and addition of a solution phase trichloroacetimidate donor allows entry to a polymer-bound disaccha-ride. Again, repetitive manipulation of the orthogonal protecting groups allows entry to larger polymer-bound systems. Once the synthesis is complete, the oligosaccharide is cleaved from the polymer. The acetate group is also employed to exert an effect on the stereochemistry of the newly formed anomeric bond due to neighbouring group participation.

In this synthesis, reaction conditions have been optimised so that each glycosylation reaction is virtually quantitative. This is essential if the reactions are to be performed successfully on a polymer support. Monitoring polymer-bound syntheses is generally very difficult and time-consuming since routine analytical techniques such as thin layer chromatography (TLC) cannot be applied to polymer-bound species. We

An acetate group is employed as a temporary orthogonal protecting group for the C-2 hydroxyl group.

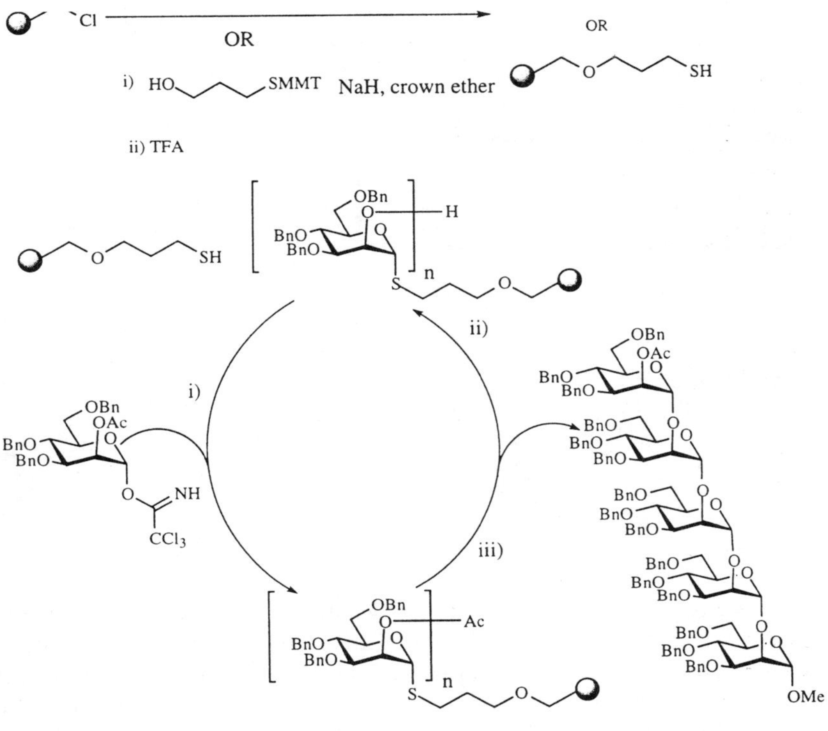

i) 3 equivs of trichloroacetimidate donor, 0.3 equivs of TMSOTf; ii) NaOMe; iii) NBS (4eq.), THF.MeOH, 9:1

Figure 7.13

will see later in the chapter that new techniques are being developed to overcome this limitation, but in this strategy the researchers resorted to cleaving small quantities of the products from the polymer to monitor the progress of the reaction. These cleaved products were then analysed using chromatographic techniques routinely used for analysis of solution-phase reactions such as thin-layer chromatography and high-performance liquid chromatography (HPLC), as well as analytical techniques such as matrix-assisted laser-desorption ionisation-time of flight (MALDI–TOF) mass spectrometry techniques.

Monitoring polymer bound syntheses is generally very difficult and time-consuming.

Use of glycals in polymer-supported syntheses

Recent work has shown that glycal donors which are linked to a polymer support, for example via the most reactive C-6 hydroxyl group, are still prone to activation with electrophiles, under conditions which have been optimised in the solution phase. In an analogous fashion to the solution phase protocol, the activated species can then be trapped by an acceptor to afford a polymer-bound oligosaccharide. If a solution-phase glycal acts as the acceptor, the extended polymer-bound glycal can be further exposed to an identical activation/trapping protocol such that an iterative process can be developed for the synthesis of extended oligosaccharides of both the linear and branched types (Figure 7.14).

Glycals also play central roles in polymer-supported syntheses of oligosaccharides.

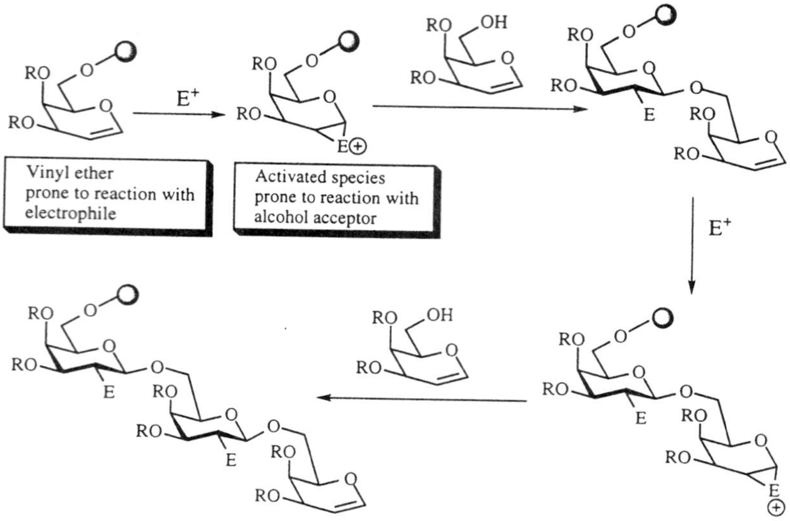

Figure 7.14

This glycal approach has proved particularly useful when a silicon-based linker is employed to attach the glycal to a commercially available polystyrene polymer and dimethyldioxirane (DMDO) is used as the activating agent, allowing formation of epoxide intermediates. The assembled oligosaccharides are easily removed from the polymer support by the use of tetrabutylammonium fluoride under acidic conditions. This effects deprotection by attacking the silyl linker, as illustrated in Figure 7.15.

Use of silyl linkers allows release of the oligosaccharide from the polymer by using TBAF.

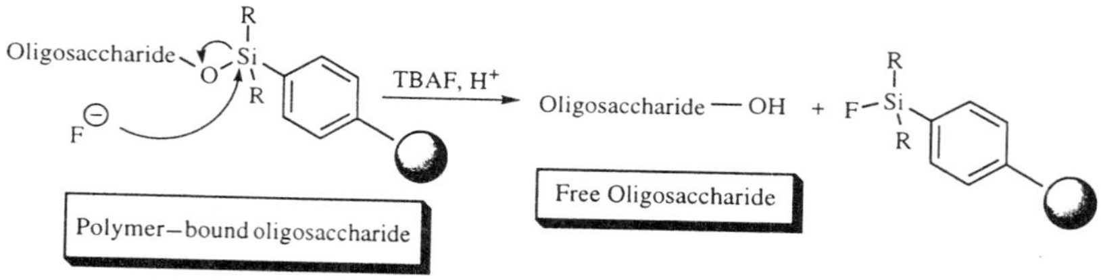

Figure 7.15

A range of acceptors with different reactivities have been employed with polymer-bound glycal donors. For example, reactive acceptors with primary hydroxyl groups, less-reactive acceptors which display a free hydroxyl group at C-3 and disaccharide acceptors have all been employed, allowing entry to both linear and branched-chain compounds. Some examples of these reactions are highlighted in Figure 7.16.

Addition of a solution-phase glycal acceptor to a polymer-bound donor allowed entry to a polymer-bound tetrasaccharide which could be further reacted with a solution-phase disaccharide acceptor to generate a polymer-bound hexasaccharide. Since any unreacted epoxide intermediate was destroyed by hydrolysis under the reaction conditions, no entities with deletions in the interior of the chain were detected. Release of the hexasaccharide from the polymer was again achieved by treatment with tetrabutylammonium fluoride under acidic conditions.

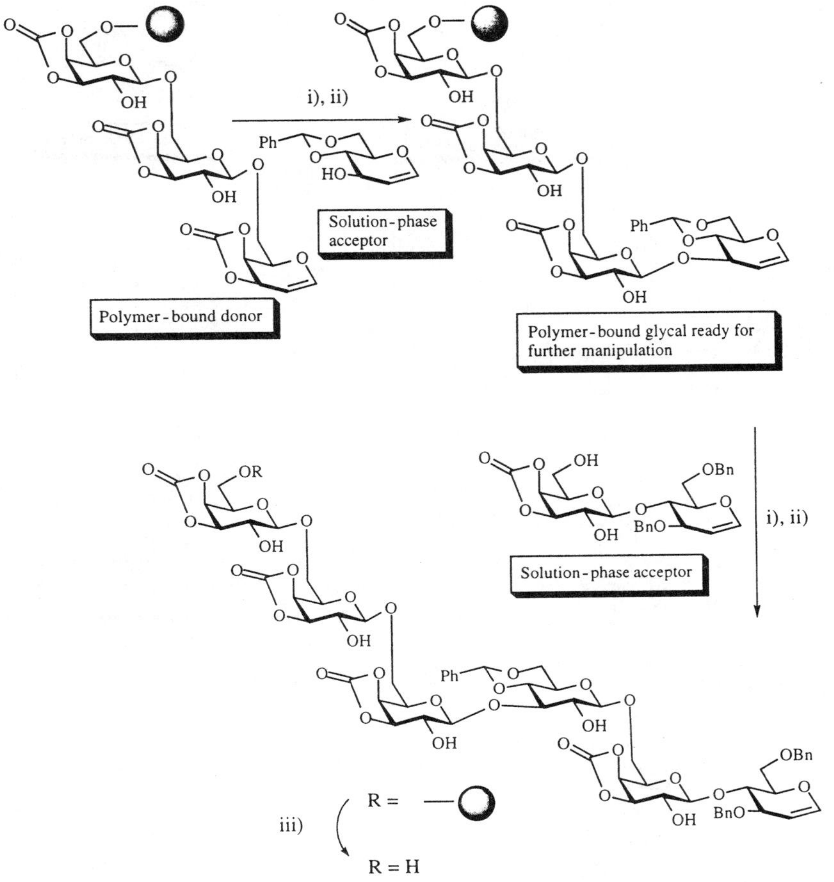

i) DMDO; ii) Acceptor, ZnCl$_2$; iii) TBAF, AcOH

Figure 7.16

Use of anomeric sulfoxides in polymer-supported syntheses

As mentioned previously, some reactions which work efficiently in the solution phase will not work as efficiently in polymer-bound strategies. This is because in the latter strategies the donors and acceptors are employed under far more dilute conditions than in solution strategies and this can lower the rate of reaction to such an extent that some processes become inefficient. Since anomeric sulfoxides are extremely reactive glycosyl donors even at low temperatures, they have proved particularly suitable as donors in polymer-bound strategies. For example, in a recent strategy, anomeric sulfoxides have been employed in conjunction with polymer-bound acceptors for the synthesis of linear oligosaccharides. The application of this protocol for the synthesis of a trisaccharide with an overall yield of 52% is outlined in Figure 7.17.

The acceptor molecule is linked to the Merrifield resin solid support via a *p*-hydroxythiophenyl linker unit. This particular linker was selected since it is stable under the reaction conditions necessary for the oligo-saccharide assembly, but is easily removed at the end of the synthesis by

In polymer-supported strategies the donors and acceptors are often employed under far more dilute conditions than in solution strategies, and this can lower the rate of reaction. Reactive sulfoxide donors can minimise this problem.

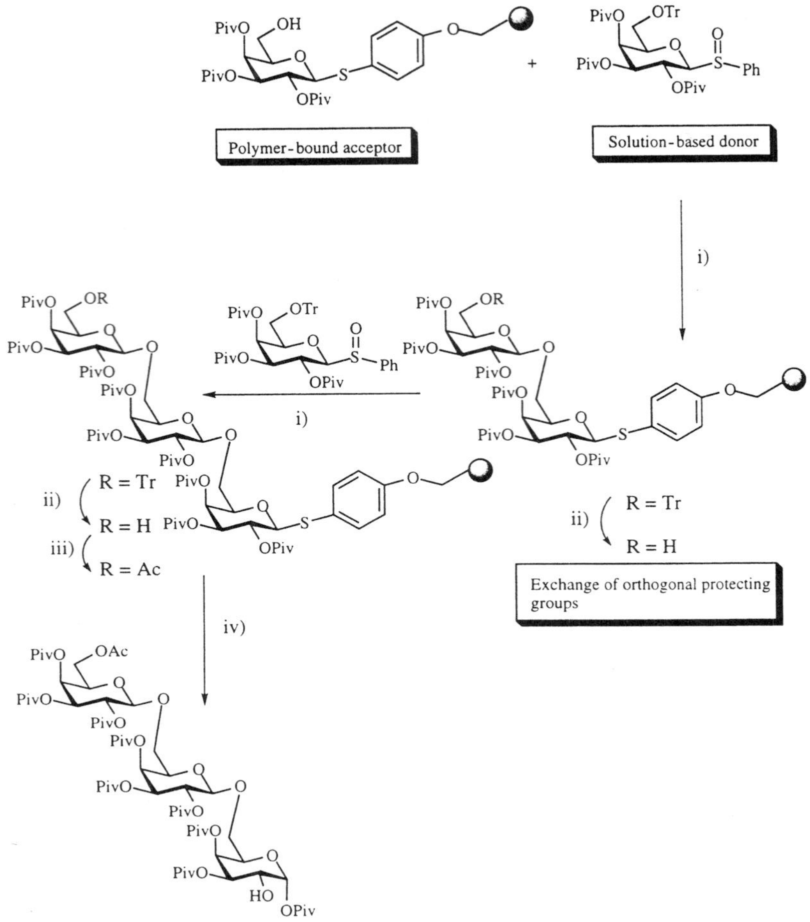

i) Tf$_2$O, 2,6-di-*tert*-butyl-4-methylpyridine; ii) CF$_3$CO$_2$H; iii) Ac$_2$O, pyr; iv) Hg(OCOCF$_3$)$_2$

Figure 7.17

treatment with mercuric trifluoroacetate. The sulfoxide donor employed was specifically designed to incorporate orthogonal protecting groups so that after the initial glycosidation reaction had been performed to afford a disaccharide on the solid support, further manipulation was possible to allow entry to branched oligosaccharides.

Use of the orthogonal glycosidation strategy in polymer-supported syntheses

The orthogonal glycosidation strategy previously introduced for solution-phase syntheses has also been applied to polymer-supported syntheses using thiomethyl and fluoride groups as the orthogonal anomeric groups. Thus a polymer-bound thiomethyl donor was reacted with a fluoro acceptor to afford a polymer-bound disaccharide which now displayed a fluoro group at the anomeric position. This could be activated by a Lewis acid and treated with a further acceptor displaying a thiomethyl group at the anomeric centre to afford the polymer-bound trisaccharide with an anomeric thiomethyl group (Figure 7.18).

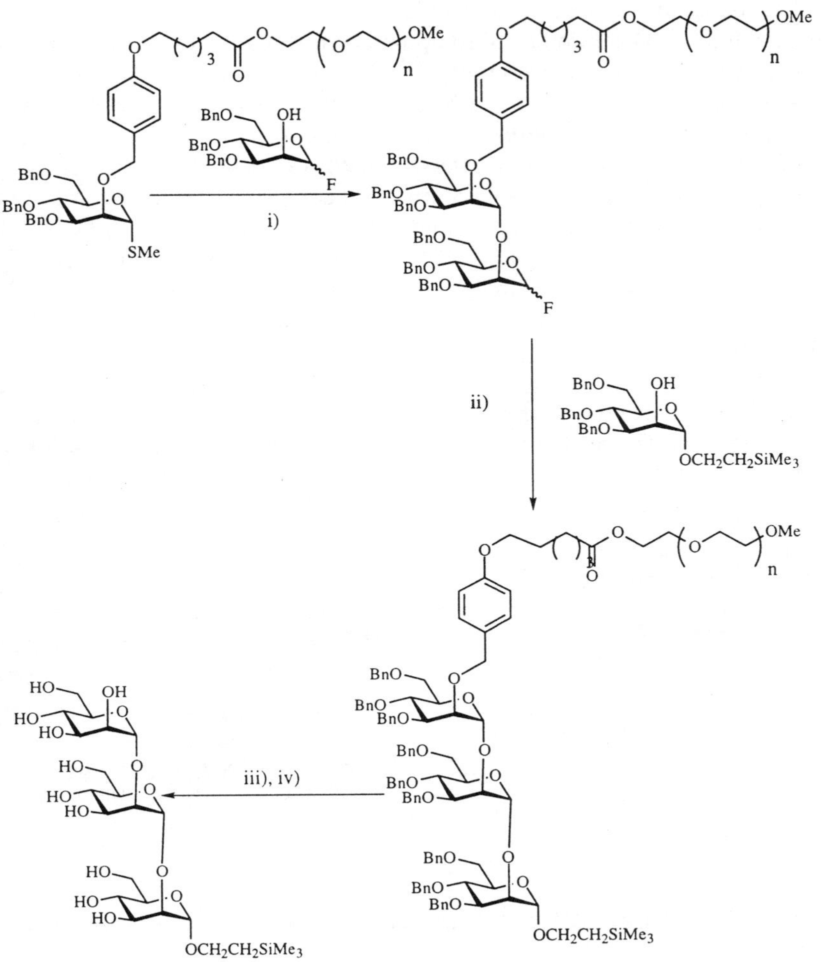

i) MeOSO$_2$CF$_3$, MeSSMe; ii) Cp$_2$HfCl, AgOSO$_2$CF$_3$; iii) NaOMe, MeOH; iv) H$_2$, Pd(OH)$_2$

Figure 7.18

The orthogonal glycosidation strategy has also been applied to polymer-supported syntheses.

This strategy hence allowed entry to an α-(1,2) linked mannotrioside in 40% overall yield from a polymer-bound donor, and a tetrasaccharide corresponding to a partially protected form of the GPI anchor, in 42% yield from a polymer-bound disaccharide. In theory this procedure could be repeated indefinitely, allowing access to large oligosaccharides.

Monitoring polymer-bound reactions

A major limitation associated with polymer-supported oligosaccharide synthesis is the difficulty in characterising reaction products or intermediates formed during the strategy whilst they are still bound to the polymer. Normally, cleavage of a small amount of product from a small portion of the polymer is required so that the cleaved product can then be analysed using classical techniques. A non-destructive analytical method which would allow the reaction to be monitored whilst the intermediates and products were still bound to the polymer support would offer many advantages. For example, it would be easier to determine the reaction times which would afford optimum yields without reducing reaction yields using destructive techniques. In this context, new NMR techniques have recently been developed which allow n.m.r. analysis of compounds bound to solvent-swollen resins. However, the quality of the spectra, particularly the breadth of the peaks obtained, varies according to the type of resin used. A Magic Angle Spinning technique using high-resolution probes (HR-MAS) has been found to afford better ^{1}H and ^{13}C NMR spectra for small resin-bound compounds. This approach has recently been extended to allow analysis of a Merrifield-bound trisaccharide glycal and this is the largest polymer-bound oligosaccharide analysed by HR-MAS to date.

Although the lines obtained in the spectra were still broader than in the resin-cleaved products, the specific peaks highlighted in Figure 7.19 allowed unequivocal confirmation of the proposed structure within a short period of time.

A Magic Angle Spinning technique using high-resolution probes (HR-MAS) has been found to afford better ^{1}H and ^{13}C NMR spectra for small resin-bound compounds. This has improved the analysis of polymer-bound reaction intermediates and products.

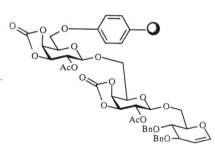

Key ^{1}H n.m.r features

Glucal 1-H 6.4 p.p.m
Acetyl C$\underline{H}_3$ approx. 2.10 p.p.m
Two PhC$\underline{H}_2$ 4 doublets at 4.83, 4.69, 4.64, 4.55 p.p.r
^{i}Pr 0.80 – 1.20 p.p.m

Key ^{13}C n.m.r features

Two acetyl C=O at 169.3 and 169.1 p.p.m
Two carbonate C=O at 154.3 and 153.6 p.p.m
Two anomeric C and Glycal C-2 at 90 – 100 p.p.m

Figure 7.19

Polymer-supported oligosaccharide synthesis employing enzymes

Many advantages offered by enzymes for the solution-phase syntheses of oligosaccharides, including their ability to initiate highly regio- and stereo-selective reactions without employing protecting groups, have recently been successfully employed with polymer-supported strategies. Recent investigations in these areas have allowed the syntheses of complex carbohydrates of biological importance, including Lewis[a] and sialyl Lewis[x]. For the former target, a disaccharide acceptor is bound to Sepharose via a disulfide linkage to afford a disaccharide which is then reacted with GDP–fucose via fucosyltransferase to afford the polymer-bound Lewis[a] trisaccharide. This is released from the polymer via treatment with mercaptoethanol to afford the trisaccharide in an overall yield of 68% (Figure 7.20).

Although such polymer-supported enzyme syntheses are in their infancy, the fact that more and more enzymes are becoming available for use in enzymatic syntheses suggests that the number of polymer-supported strategies employing enzymes will also increase over the next few years.

In this chapter we have discussed some alternative methods for synthesising oligosaccharides which offer a number of advantages compared

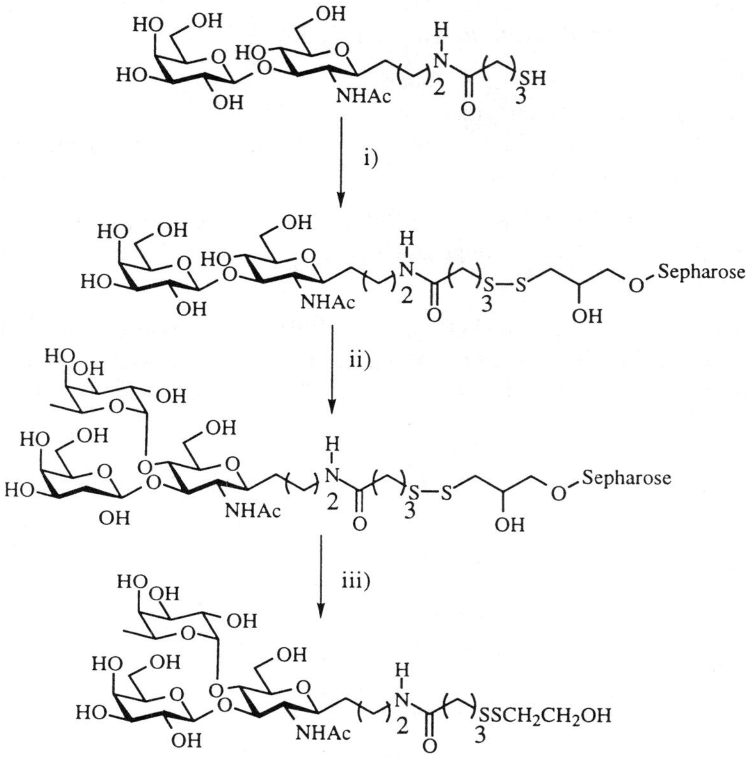

i) Activated thiopropyl–Sepharose; ii) Fucosyltransferase, GDP–Fucose; iii) 2-Mercaptoethanol, 68% overall yield

Many advantages offered by enzymes for the solution-phase syntheses of oligosaccharides have recently been successfully employed with polymer-supported strategies.

Figure 7.20

with purely chemical strategies. Although the polymer-supported synthesis of oligosaccharides is still a relatively new area of research, some of the recent advances highlighted in this chapter offer hope that polymer-supported synthesis of oligosaccharides may become more practical in the near future.

Further reading

1. Wong, C.-H., Halcomb, R.L., Ichikawa, Y., Kajimoto, T. (1995). Enzymes in organic synthesis: application to the problems of carbohydrate recognition (Part 1). *Angewandte Chemie International Edition in English*, **34**, 412–432.

2. Wong, C.-H., Halcomb, R.L., Ichikawa, Y., Kajimoto, T. (1995). Enzymes in organic synthesis: application to the problems of carbohydrate recognition (Part 2). *Angewandte Chemie International Edition in English*, **34**, 521–546.

3. David, S., Augé, C., Gautheron, C. (1991). Enzymatic methods in preparative carbohydrate chemistry. *Advances in Carbohydrate Chemistry and Biochemistry*, **49**, 176–238.

4. Augé, C., Crout, D.H.G. (1997). Chemoenzymatic synthesis of carbo-hydrates. *Carbohydrate Research*, **305**, 307–312.

5. Watt, G.M., Revers, L., Webberly, M.C., Wilson, I.B.H., Flitsch, S.L. (1998). The chemoenzymatic synthesis of the core trisaccharide of *N*-linked oligosaccharides using a recombinant β-mannosyltransferase. *Carbohydrate Research*, **305**, 533–541.

5. Osborn, H.M.I., Khan, T.H. (1999). Recent developments in polymer supported synthesis of oligosaccharides and glycopeptides. *Tetrahedron*, **55**, 1807–1851.

6. Seeberger, P.H., Danishefsky, S.J. (1998). Solid-phase synthesis of oligosaccharides and glycoconjugates by the glycal assembly method: a five year retrospective. *Accounts of Chemical Research*, **31**, 685–695.

Index